国外优秀数学著作
原　版　系　列

基于不确定静态和动态问题解的仿射算术

Affine Arithmetic Based Solution of Uncertain Static and Dynamic Problems

（英文）

［印］斯内哈希什·查克拉弗蒂（Snehashish Chakraverty）
［印］绍达米尼·劳特（Saudamini Rout）
著

HITP
哈爾濱工業大學出版社
HARBIN INSTITUTE OF TECHNOLOGY PRESS

黑版贸审字 08-2021-033 号

图书在版编目(CIP)数据

基于不确定静态和动态问题解的仿射算术=Affine Arithmetic Based Solution of Uncertain Static and Dynamic Problems:英文/(印)斯内哈希什・查克拉弗蒂(Snehashish Chakraverty),(印)绍达米尼・劳特(Saudamini Rout)著.—哈尔滨:哈尔滨工业大学出版社,2023.3

ISBN 978-7-5767-0731-1

Ⅰ.①基… Ⅱ.①斯… ②绍… Ⅲ.①模糊数学-英文 Ⅳ.①O159

中国国家版本馆 CIP 数据核字(2023)第 050827 号

JIYU BUQUEDING JINGTAI HE DONGTAI WENTI JIE DE FANGSHE SUANSHU

策划编辑　刘培杰　杜莹雪
责任编辑　关虹玲　张嘉芮
封面设计　孙茵艾
出版发行　哈尔滨工业大学出版社
社　　址　哈尔滨市南岗区复华四道街 10 号　邮编 150006
传　　真　0451-86414749
网　　址　http://hitpress. hit. edu. cn
印　　刷　哈尔滨市工大节能印刷厂
开　　本　787 mm×1 092 mm　1/16　印张 11.25　字数 182 千字
版　　次　2023 年 3 月第 1 版　2023 年 3 月第 1 次印刷
书　　号　ISBN 978-7-5767-0731-1
定　　价　38.00 元

(如因印装质量问题影响阅读,我社负责调换)

Contents

Preface

Static and dynamic problems may be modeled by systems of simultaneous equations, eigenvalue problems, differential equations, and integral equations. Generally, the material and geometric properties (or involved parameters) of these problems are considered to be in the form of crisp values. Uncertainty is an inseparable companion of almost every measurement, and occurrence of uncertainty is a must while dealing with real-world problems. Thus, in actual practice, due to maintenance, measurement, or experiment-induced errors, the material properties may be uncertain or vague. There are three classes of uncertain models viz. probabilistic or statistical methods, interval analysis, and fuzzy set theory. In the probabilistic approach, the uncertain parameters are considered as random variables, whereas, in the interval and fuzzy computations, the parameters are treated as closed intervals of real line $\mathbb{R}$ and fuzzy numbers, respectively. Traditionally, probabilistic or statistical approaches are used to handle the uncertainties and vagueness. Unfortunately, probabilistic methods might not deliver reliable results at the required precision without a sufficient amount of experimental data about the involved parameters. As such, interval analysis and fuzzy set theory have become powerful tools for many applications in recent decades. In fuzzy set theory and interval analysis, the uncertain parameters are expressed by fuzzy and interval variables, such as fuzzy and interval numbers, vectors, or fuzzy and interval matrices. Further, the fuzzy numbers can be parameterized and transformed into a family of intervals. The main obstacle while handling the uncertainties with interval analysis is its dependency problem. The interval arithmetic assumes that all the operands of a computation vary independently over their ranges while performing any interval operations. But, when the operands are partially dependent, interval arithmetic may result in a larger width of the interval solutions than the exact range. In this respect, affine arithmetic (AA) may be one of the recent developments to handle the uncertainties in a different manner which may be useful to overcome the dependency problem and compute better enclosures for the solutions. Further, as per our aim, static and dynamic problems turn into linear/nonlinear systems of equations and eigenvalue problems, respectively. As such, affine arithmetic based methods are proposed/developed to solve the above problems and handle the uncertainty with ease.

In this regard, this book will start with the basic premises of interval and fuzzy arithmetic. Then it will address the dependency problem in standard interval arithmetic and the need for affine arithmetic. Exhaustive affine arithmetic operations and properties will be included with fuzzy-affine arithmetic. It may be noted that static problems with uncertainties lead to uncertain systems of linear equations, and uncertain dynamic problems turn into uncertain linear and nonlinear eigenvalue problems.

As such, this book includes newly developed efficient methods to handle the described problems based on affine and interval/fuzzy approach. Various illustrative examples with respect to the static and dynamic problems of structures are investigated to show the reliability and efficacy of all the developed approaches. The authors assume that readers have essential knowledge of calculus, real analysis, linear algebra, numerical analysis, and differential equations.

This book consists of seven chapters. Chapter 1 addresses the detailed introduction and literature studies of the titled problem. Basic definitions, terminologies, and properties related to interval analysis and fuzzy set theory are included in Chapter 2. In Chapter 3, the reason behind the interval dependency problem and a detailed explanation of affine arithmetic is discussed. New fuzzy-affine arithmetic is proposed in Chapter 4 to efficiently handle real-life problems with uncertainties in the form of fuzzy numbers. In Chapter 5, an investigation of uncertain static problems has been incorporated, which may lead to the interval and/or fuzzy systems of linear equations. Dynamic analysis with uncertain material and geometric properties of various practical problems may be modeled through interval and/or fuzzy eigenvalue problems. As such, different approaches for solving uncertain linear and nonlinear dynamic problems are presented in Chapter 6 and Chapter 7, respectively. Throughout the book, uncertainty is taken in the form of closed intervals, and fuzziness is considered in terms of triangular and trapezoidal fuzzy numbers.

This book is an attempt to rigorously address the study of static and dynamic problems from various scientific and engineering fields in uncertain environments. The authors believe that this book will be helpful to undergraduates, graduates, researchers, industry, faculties, and others throughout the globe.

Snehashish Chakraverty and Saudamini Rout
March 2020

Acknowledgments

It has been great to explore the vast area of **Affine Arithmetic Based Solution of Uncertain Static and Dynamic Problems**. The authors hope it will benefit readers who may want to start learning about the subject by reading this challenging book. The authors have completed this book amid moments of both immense hardship and joy while being vastly supported by wonderful people who have always been on their side.

As such, the first author would like to thank his parents for their blessings and motivation. Next, he would like to thank his wife, Shewli, and daughters, Shreyati and Susprihaa, for their support and inspiration during this project. The support of all the Ph.D. students of the first author as well as the NIT Rourkela administration and facilities are also gratefully acknowledged.

The second author is extremely grateful to her parents, Mr. Siba Prasad Rout and Mrs. Swapna Rani Rout, for their love, understanding, prayers, and constant support during her ongoing research work for this book. Gratitude is also extended to all her family members, relatives, and respected teachers for their love, care, motivation, and blessings throughout the writing of this book. Also, she would like to thank her friends, who have been constantly there for supporting her mentally. She is greatly indebted to all her fellow lab mates for their constant support.

Also, the second author would like to thank NIT Rourkela for giving her an opportunity to work with Prof. Snehashish Chakraverty, first author of the book, who has been very supportive and cooperative in all respects. His immense knowledge, guidance, and motivation have been a driving force to complete this book. He has always been very generous in drawing the outlines of good research and clear presentation of it.

Further, the authors sincerely acknowledge the reviewers for their fruitful suggestions. All the authors appreciate the support and help of the whole team at Morgan & Claypool Publishers. Finally, the authors are greatly indebted to all the authors/researchers mentioned in the reference sections given at the end of each chapter.

Snehashish Chakraverty and Saudamini Rout
March 2020

CHAPTER 1

Introduction

In general, governing differential equations of static and dynamic problems from several fields of engineering and science viz. structural mechanics, acoustic systems, simulation of electrical circuits, modeling micro-electronic mechanical systems, fluid mechanics, and signal processing lead to systems of linear equations and eigenvalue problems. These may be linear (standard and generalized) and nonlinear eigenvalue problems. The mathematical models of structural systems are governed by the following differential equation:

$$M\ddot{s}(t) + C\dot{s}(t) + Ks(t) = f(t), \tag{1.1}$$

where M, C, and K are the respective mass, damping factor, and stiffness matrices. Further, $s(t)$ is the displacement vector and $f(t)$ is the external load vector.

The governing differential equation (1.1) converts to the following linear system under static conditions.

$$Ks = f. \tag{1.2}$$

Further, under a dynamic case, the governing differential equation may be modeled to eigenvalue problems. Particularly, for an undamped system, Eq. (1.1) reduced to

$$M\ddot{s}(t) + Ks(t) = f(t), \tag{1.3}$$

which may lead to a generalized eigenvalue problem as

$$Kx = \lambda Mx. \tag{1.4}$$

Similarly, the dynamic analysis of a damped structural system transforms into a nonlinear eigenvalue problem (particularly, a quadratic eigenvalue problem) as

$$\left(M\lambda^2 + C\lambda + K\right)x = 0. \tag{1.5}$$

In Eqs. (1.4) and (1.5), λ is the eigenvalue (natural frequency of the structural system) and x is the corresponding eigenvector of the system known as vibration characteristics.

For simplicity and easy computations, all the involved parameters and variables are usually considered as crisp (or deterministic or exact). But, as a practical matter, due to the uncertain environment, one may have imprecise, incomplete, insufficient, or vague information about the parameters because of several errors. Traditionally, such uncertainty or vagueness may be modeled through a probabilistic or statistical approach. But a large amount of experimental data is required for these traditional probabilistic and statistical approaches. Without a sufficient amount of experimental data, these methods may not deliver reliable results at the required precision. Therefore, intervals and/or fuzzy numbers may be used to handle uncertain and vague parameters when there is an insufficient amount of data available.

The "dependency problem" in interval analysis is a major hurdle while handling uncertainties with interval computations. Standard interval arithmetic treats all its operands independently over their ranges. But in cases where the operands partially depend upon each other, interval computation results in comparatively wide intervals rather than the exact solution. This is the main reason behind the interval dependency problem. Due to this scenario, for complex computations, the overestimation of range increases with each iteration.

In this regard, affine arithmetic was recently developed to handle uncertainties efficiently. Affine arithmetic is a self-validated numerical model that deals with the dependency problem in the case of standard interval computations. The major advantage of affine arithmetic is that it keeps track of the first-order correlations between all the inputted and computed values. These first-order correlations are exploited automatically in every primitive operation. For this reason, affine arithmetic may be able to overcome the interval dependency problem and produce much better interval estimates than the standard interval arithmetic. Furthermore, affine arithmetic may also implicitly be able to provide a geometric representation of the joint range of all the related quantities, which may be exploited to increase the reliability of the interval methods that are used in various theoretical as well as applied research.

It already has been mentioned that affine arithmetic based computations have various advantages over standard interval arithmetic. However, the literature review reveals that very few investigations are available to handle the titled problem through affine arithmetic. As such, the main goal of the book is to handle various static and dynamic problems by different interval and/or fuzzy computations and further by infusing affine arithmetic.

Previous important studies related to the said problem are included in the following section, categorized into different heads such as:

- interval analysis
- fuzzy set theory
- affine arithmetic
- uncertain static problems
- uncertain linear dynamic problems
- uncertain nonlinear dynamic problems

1.1 LITERATURE REVIEW

In this section, first, a few related studies of interval and fuzzy arithmetic are included. Accordingly, some research work on affine arithmetic has been incorporated. Further, literature emphasizing static, linear and nonlinear dynamic problems is discussed.

1.1.1 INTERVAL ANALYSIS

The basic concepts and terminologies of standard interval arithmetic were first introduced by R. E. Moore in the early 1960s. As such, a few related works of literature are given here. Moore (1962) [61] discussed interval arithmetic and automatic error analysis in the case of digital computing. Further, the detailed methods and applications of interval analysis were included in Moore (1979) [62] and Moore et al. (2009) [63]. Alefeld and Herzberger (2012) [6] introduced the concept of interval computations. Applications of interval analysis for parameter and state estimation in engineering problems viz. robust control and robotics were discussed in Jaulin et al. (2001) [42]. Krämer (2006) [47] and Hansen (1975) [35] proposed generalized intervals and their arithmetic with the dependency problem.

1.1.2 FUZZY SET THEORY

Few literature studies related to the basic concepts and properties of fuzzy set theory have been discussed in this section. Zadeh (1965) [105] first introduced the concepts of fuzzy sets and fuzzy numbers in 1965. In his work, Zadeh generalized the classical sets with characteristic functions, which vary over [0, 1]. Fuzzy set theory is an important tool to handle uncertain parameters. In this regard, excellent books have been written by different authors viz., Dubois (1980) [30], Kaufmann and Gupta (1988) [44], Zimmermann (2011) [107], Hanss (2005) [36], Zadeh et al. (2014) [106], Chakraverty et al. (2016) [24], Chakraverty and Perera (2018) [23], and others.

1.1.3 AFFINE ARITHMETIC

Standard interval arithmetics' underlying assumption about the independence of all the operands is responsible for the interval overestimation problem (or decency problem) where it overestimates the width of the resulting interval solutions. In this regard, affine arithmetic proves itself as an efficient tool to handle the overestimation problem and results in comparatively tighter enclosures. The concept of affine arithmetic and its applications in computer graphics were first introduced by Comba and Stolfi (1993) [25]. After a few years, Stolfi and De Figueiredo (2003) [92] illustrated the overestimation problem in the case of standard interval arithmetic and how affine arithmetic can overcome it. Further, the concepts, properties and several applications of affine arithmetic were also discussed by De Figueiredo and Stolfi (2004) [28]. Miyajima and Kashiwagi (2004) [60] proposed a dividing method by utilizing the best multiplication in affine arithmetic. Moreover, Akhmerov (2005) [5] developed an interval-affine Gaussian algorithm for constrained systems. A direct method for solving parametric interval linear systems with non-affine dependencies was demonstrated by Skalna (2009) [88]. Rump and Kashiwagi (2015) [84] discussed the improvements and implementations of affine arithmetic. Skalna and Hladík (2017) [89] developed a new algorithm for Chebyshev minimum-error multiplication of reduced affine forms. An optimization model based on improved affine arithmetic for interval power flow analysis was developed by Xu et al. (2016) [103]. Furthermore, Adusumilli and Kumar (2018) [4] studied the modified affine arithmetic based continuation power flow analysis for voltage stability assessment under uncertainty. An affine arithmetic-based energy management system for isolated microgrids was illustrated by Romero-Quete and Cañizares (2018) [81]. Wang et al. (2018) [98] discussed an affine arithmetic-based multi-objective optimization method for energy storage systems operating in active distribution networks with uncertainties.

1.1.4 UNCERTAIN STATIC PROBLEMS

Under static conditions, the governing differential equations of various science and engineering problems may lead to systems of linear equations. Depending upon the parameters of the uncertain static problems, this instance may be referred to as an interval or fuzzy linear system of equations. As such, a few related works are presented here.

Crisp System of Linear Equations (CSLE)

There are various well-known methods to solve CSLE. As such, good literature related to this is already available. A few important works of literature are discussed here. Tanabe (1971) [95] proposed a projection method to solve a singular CSLE and discussed its applications. The solution of an overdetermined CSLE was described by Barrodale and Phillips (1975) [13]. Horn and Johnson (1985) [41], Strang (1993) [93], and Meyer (2000) [59] described other techniques to handle CSLEs. Further, the homotopy perturbation method (HPM) has also been used to solve the CSLE. Yusufoğlu (2009) [104] proposed an improvement to the HPM for solving CSLEs. To find the solution of the CSLE, Keramati (2009) [45] developed a method viz. He's HPM. A modified HPM was discussed by Noor et al. (2013) [67] for solving the CSLE.

Interval System of Linear Equations (ISLE)

In this section, some literature related to the solution of an ISLE is provided. The theories and applications of CSLE with detailed techniques to find solution bounds are found in Neumaier (1986) [64] and Neumaier (1990) [65]. Several methods and their comparative study for solving CSLE were illustrated by Ning and Kearfott (1997) [66]. For computing the static bounds of displacement (solution of CSLE), Qiu and Elishakoff (1998) [72] investigated the subinterval methods and interval perturbation methods. Furthermore, McWilliam (2001) [57] proposed two new methods based on the monotonicity of displacements and modified interval perturbation analysis to solve CSLE. Hladík (2012) [37] discussed the enclosures for the solution set of parametric interval linear systems. Chakraverty et al. (2017) [20] discussed the formal solution of an interval system of linear equations with an application in static responses of structures with interval forces. Chakraverty et al. (2017) [21] presented a sign function approach to solve an algebraically interval system of linear equations for nonnegative solutions. Karunakar and Chakraverty (2018) [43] studied the interval solution of fully interval linear systems of equations by employing the criteria of tolerable solution.

Fuzzy System of Linear Equations (FSLE)

Research work related to computing the solution of FSLE, is included in this section. Dehghan et al. (2006) [29] described several new methods based on the Dubois and Prade (1980) [31] approach for finding the nonnegative solution vector of FSLE in which the constituent matrices are the fuzzy coefficient matrix and corresponding right-hand side fuzzy vector. Further, Allahviranloo (2004) [7] proposed several numerical methods to find the solution of the FSLE. A successive over-relaxation iterative method and an Adomian decomposition method were also proposed by Allahviranloo in Allahviranloo (2005a) [8] and Allahviranloo (2005b) [9], respectively. A block Jacobi two-stage method with Gauss–Sidel inner iterations was described by Allahviranloo et al. (2006) [10] for fuzzy linear systems. Moreover, Abbasbandy et al. (2005) [3] proposed the conjugate gradient method to evaluate the solution of the FSLE, in which the system is considered to be a fuzzy symmetric positive definite system. Also, the LU decomposition method was illustrated by Abbasbandy et al. (2006) [1] for solving the FSLE. Abbasbandy and Jafarian (2006) [2] illustrated the steepest descent method to handle fuzzy linear systems. The fuzzy complex system of linear equations was adapted to circuit analysis by Rahgooy et al. (2009) [76]. Das and Chakraverty (2012) [27] proposed a new numerical algorithm for solving fuzzy and interval system of linear equations. Based on the single and double parametric form of fuzzy numbers, Behera and Chakraverty (2015) [15] proposed a new approach to solve fully fuzzy systems of linear equations.

1.1.5 UNCERTAIN LINEAR DYNAMIC PROBLEMS

The dynamic analysis of science and engineering problems may lead to linear eigenvalue problems (LEPs), such as the generalized eigenvalue problem (GEP) and the standard eigenvalue problem (SEP). Depending upon the uncertain parameters, an LEP may be classified as interval and/or fuzzy GEP (or SEP). In this regard, a few literatures studies are discussed below.

Crisp Generalized/Standard Eigenvalue Problem (CGEP/CSEP)

There are various well-known methods to solve CGEP or CSEP. Wilkinson (1965) [101] studied the concept and properties of the algebraic eigenvalue problem. Bathe and Wilson (1973) [14] discussed different methods to find solutions for eigenvalue problems from structural mechanics. Reduction technique for a band-symmetric CGEP was illustrated by Crawford (1973) [26]. Stewart (1975) [91] proposed a Gershgorin theory for CGEP in the form $Ax = \lambda Bx$. A balancing approach to solve CGEP was introduced by Ward (1981) [99]. Bunse-

Gerstner (1984) [17] described an algorithm for the solution of CGEP in which the coefficient matrices of the problem are taken as symmetric. Lanczos algorithm was applied to handle the unsymmetric CGEP by Rajakumar and Rogers (1991) [78].

Interval Generalized/Standard Eigenvalue Problem (IGEP/ISEP)

Different works are available for the solution of interval linear eigenvalue problems viz. IGEP ($[G][x] = [\lambda][H][x]$) and ISEP ($[S][x] = [\lambda][x]$). Qiu et al. (1996) [70] computed the eigenvalue bounds for structures with the interval description of uncertain but non-random parameters. The eigenvalue bounds of interval matrices also were evaluated by Rohn (1998) [80]. Qiu et al. (1995) [71] proposed the rayleigh quotient iteration method for calculating the eigenvalue bounds of structures having bounded uncertain parameters. Further, the modal analysis of structures and computing eigenvalue bounds having uncertain but bounded parameters with the help of interval analysis were studied by Qiu et al. (2005) [75] and Sim et al. (2007) [87]. Qiu et al. (2001) [73] described an approximate method to solve ISEPs of real non-symmetric interval matrices. Qiu and Wang (2005) [74] illustrated solution theorems for ISEP of structures with uncertain but bounded parameters. Under the general form parametric dependencies, Kolev (2006) [46] derived the outer interval solutions of IGEPs. Based on perturbation theory, uncertain but non-random eigenvalue bounds of structures were computed by Leng and He (2007) [52]. Also, Leng et al. (2008) [54] and Leng and He (2010) [53] described the computation of the real eigenvalue bounds of the real interval matrices in structural dynamics with interval parameters. Hladík and Jaulin (2011) [40] discussed the eigenvalue symmetric matrix contractor. Hladík et al. (2011) [39] proposed a filtering method for solving interval eigenvalue problems, which is based on the concept of sufficient conditions for singularity and regularity of interval matrices given by Rex and Rohn (1998) [79]. Furthermore, Hladík (2013) [38] found the eigenvalue solution bounds of the IGEP having both real and complex interval matrices. The real eigenvalue bounds of the real IGEP and real ISEP were studied by Leng (2014) [51].

Fuzzy Generalized/Standard Eigenvalue Problem (FGEP/FSEP)

Literature studies related to solving fuzzy linear eigenvalue problems such as FGEP and FSEP are discussed here. An efficient technique to find the solution of FGEPs in structural dynamics was proposed by Xia and Friswell (2014) [102]. Further, Mahato and Chakraverty (2016b) [56] extended the filtering algorithm given by Hladik et al. (2011) [39] for eigenvalue bounds of fuzzy symmetric matrices to solve FSEP. Also, the filtering algorithm for real eigenvalue bounds of both IGEP and FGEP were illustrated by Mahato and Chakraverty (2016a) [55]. Chakraverty and Behera (2014) [18] discussed the parameter identification of multi-story frame structure with uncertain dynamic data, which may lead to FGEP. Uncertain static and dynamic analysis of structural systems with imprecise parameters was described by Chakraverty and Behera (2017) [19].

1.1.6 UNCERTAIN NONLINEAR DYNAMIC PROBLEMS

Nonlinear dynamic problems from various fields of science and engineering lead to nonlinear eigenvalue problems (NEPs). Similarly, these are also categorized into the interval and/or fuzzy NEPs. Some literature studies on these problems are given below.

Crisp Nonlinear Eigenvalue Problem (CNEP)

There are various methods for the solution of CNEP, such as Newton's method, linearization method, improved Newton method, second-order Arnoldi method, backward error analysis technique using linearization, Lanczos algorithm, block Newton method, and so on. As such, investigations have been carried out by several researchers to handle CNEPs. Basic theories and properties of matrix structural analysis were presented by Przemieniecki (1968) [69]. Further, Gohberg et al. (1982) [34] discussed various concepts and properties of matrix polynomials. Rajakumar (1993) [77] developed the Lanczos algorithm for the solution of quadratic eigenvalue problems (QEPs) in engineering applications. The backward error analysis of polynomial eigenvalue problems (a particular type of CNEP in which the nonlinear matrix-valued function is considered as a matrix-valued polynomial function) was explained by Tisseur (2000) [96]. Tisseur and Meerbergen (2001) [97] illustrated a review work on solving the crisp quadratic eigenvalue problem (CQEP) (a nonlinear eigenvalue problem of order two) using different numerical techniques. Their work is based upon the linearization of the CNEP into a generalized/standard eigenvalue problem. Mehrmann and Watkins (2002) [58] studied the linearization procedure to solve polynomial eigenvalue problems with the Hamiltonian structure. The structured eigenvalue method for the computation of corner singularities in 3D anisotropic elastic structures was studied by Apel et al. (2002) [11]. Further, Bai and Su (2005) [12] proposed a second-order Arnoldi method (SOAR) to solve CQEP. The study of a special form of CNEP known as rational eigenvalue problems and its solution via linearization technique was presented by Su and Bai (2011) [94]. Wetherhold and Padliya (2014) [100] designed different aspects of nonlinear vibration analysis of rectangular orthotropic membranes. Backward error analysis of polynomial eigenvalue problems solved by linearization was studied by Lawrence et al. (2016) [50]. Solov'ev and Solov'ev (2018) [90] developed a finite element approximation of the minimal eigenvalue of a nonlinear eigenvalue problem. The nonlinear eigenvalue problem was solved by Saad et al. (2019) [85] using a rational approximation method. The linearized and nonlinearized solutions of nonlinear multiparameter eigenvalue problems were discussed by Kurseeva et al. (2019) [49]. Bender et al. (2019) [16] studied nonlinear eigenvalue problems for generalized Painlevé equations. Furthermore, the Newton method is widely used to solve CNEP. As such, Kressner (2009) [48] developed a block Newton method for the solution of CNEPs. An improved Newton method was also proposed by Fazeli and Rabiei (2016) [32] to solve CNEPs. Moreover, Gao et al. (2009) [33] described the solution of aclass of CNEPs by Newton's method.

Interval Nonlinear Eigenvalue Problem (INEP)

There exist very few literature studies for the solution of INEPs. Chakraverty and Mahato (2018) [22] solved INEP for damped spring-mass systems using two methods viz. the linear sufficient regularity perturbation (LSRP) and direct sufficient regularity perturbation (DSRP) methods. Further, by using the computations of standard interval arithmetic, Sadangi (2013) [86] solved the INEPs. The solution to the INEP of the structural systems in the uncertain environment was investigated by Rout and Chakraverty (2020) [83].

Fuzzy Nonlinear Eigenvalue Problem (FNEP)

To the best of our knowledge, no work can be found regarding solving FNEP, although, Rout and Chakraverty (2019) [82] studied FNEPs for damped spring-mass structural systems and solved them by using a novel fuzzy-affine approach.

1.2 REFERENCES

[1] Abbasbandy, S., Ezzati, R., and Jafarian, A., 2006. LU decomposition method for solving fuzzy system of linear equations. *Applied Mathematics and Computation*, 172(1):633–643. DOI: 10.1016/j.amc.2005.02.018. 5

[2] Abbasbandy, S. and Jafarian, A., 2006. Steepest descent method for system of fuzzy linear equations. *Applied Mathematics and Computation*, 175(1):823–833. DOI: 10.1016/j.amc.2005.07.036. 5

[3] Abbasbandy, S., Jafarian, A., and Ezzati, R., 2005. Conjugate gradient method for fuzzy symmetric positive definite system of linear equations. *Applied Mathematics and Computation*, 171(2):1184–1191. DOI: 10.1016/j.amc.2005.01.110. 5

[4] Adusumilli, B. S. and Kumar, B. K., 2018. Modified affine arithmetic based continuation power flow analysis for voltage stability assessment under uncertainty. *IET Generation, Transmission and Distribution*, 12(18):4225–4232. DOI: 10.1049/iet-gtd.2018.5479. 4

[5] Akhmerov, R. R., 2005. Interval-affine Gaussian algorithm for constrained systems. *Reliable Computing*, 11(5):323–341. DOI: 10.1007/s11155-005-0040-5. 3

[6] Alefeld, G. and Herzberger, J., 2012. *Introduction to Interval Computation*. Academic Press, London. DOI: 10.1016/C2009-0-21898-8. 3

[7] Allahviranloo, T., 2004. Numerical methods for fuzzy system of linear equations. *Applied Mathematics and Computation*, 155(2):493–502. DOI: 10.1016/s0096-3003(03)00793-8. 5

[8] Allahviranloo, T., 2005a. Successive over relaxation iterative method for fuzzy system of linear equations. *Applied Mathematics and Computation*, 162(1):189–196. DOI: 10.1016/j.amc.2003.12.085. 5

[9] Allahviranloo, T., 2005b. The Adomian decomposition method for fuzzy system of linear equations. *Applied Mathematics and Computation*, 163(2):553–563. DOI: 10.1016/j.amc.2004.02.020. 5

[10] Allahviranloo, T., Ahmady, E., Ahmady, N., and Alketaby, K. S., 2006. Block Jacobi two-stage method with Gauss–Sidel inner iterations for fuzzy system of linear equations. *Applied Mathematics and Computation*, 175(2):1217–1228. DOI: 10.1016/j.amc.2005.08.047. 5

[11] Apel, T., Mehrmann, V., and Watkins, D., 2002. Structured eigenvalue methods for the computation of corner singularities in 3D anisotropic elastic structures. *Computer Methods in Applied Mechanics and Engineering*, 191(39–40):4459–4473. DOI: 10.1016/s0045-7825(02)00390-0. 7

[12] Bai, Z. and Su, Y., 2005. SOAR: A second-order Arnoldi method for the solution of the quadratic eigenvalue problem. *SIAM Journal on Matrix Analysis and Applications*, 26(3):640–659. DOI: 10.1137/s0895479803438523. 7

[13] Barrodale, I. and Phillips, C., 1975. Algorithm 495: Solution of an overdetermined system of linear equations in the Chebychev norm [F4]. *ACM Transactions on Mathematical Software (TOMS)*, 1(3):264–270. DOI: 10.1145/355644.355651. 4

[14] Bathe, K. J. and Wilson, E. L., 1973. Solution methods for eigenvalue problems in structural mechanics. *International Journal for Numerical Methods in Engineering*, 6(2):213–226. DOI: 10.1002/nme.1620060207. 5

[15] Behera, D. and Chakraverty, S., 2015. New approach to solve fully fuzzy system of linear equations using single and double parametric form of fuzzy numbers. *Sadhana*, 40(1):35–49. DOI: 10.1007/s12046-014-0295-9. 5

[16] Bender, C. M., Komijani, J., and Wang, Q. H., 2019. Nonlinear eigenvalue problems for generalized Painlevé equations. *Journal of Physics A: Mathematical and Theoretical*. DOI: 10.1088/1751-8121/ab2bcc. 7

[17] Bunse-Gerstner, A., 1984. An algorithm for the symmetric generalized eigenvalue problem. *Linear Algebra and its Applications*, 58:43–68. DOI: 10.1016/0024-3795(84)90203-9. 6

[18] Chakraverty, S. and Behera, D., 2014. Parameter identification of multistorey frame structure from uncertain dynamic data. *Strojniški Vestnik-Journal of Mechanical Engineering*, 60(5):331–338. DOI: 10.5545/sv-jme.2014.1832. 6

[19] Chakraverty, S. and Behera, D., 2017. Uncertain static and dynamic analysis of imprecisely defined structural systems. In *Fuzzy Systems: Concepts, Methodologies, Tools, and Applications*, pages 1–30, IGI Global. DOI: 10.4018/978-1-5225-1908-9.ch001. 6

[20] Chakraverty, S., Hladík, M., and Behera, D. 2017. Formal solution of an interval system of linear equations with an application in static responses of structures with interval forces. *Applied Mathematical Modelling*, 50:105–117. DOI: 10.1016/j.apm.2017.05.010. 4

[21] Chakraverty, S., Hladík, M., and Mahato, N. R., 2017. A sign function approach to solve algebraically interval system of linear equations for nonnegative solutions. *Fundamenta Informaticae*, 152(1):13–31. DOI: 10.3233/fi-2017-1510. 5

[22] Chakraverty, S. and Mahato, N. R., 2018. Nonlinear interval eigenvalue problems for damped spring-mass system. *Engineering Computations*, 35(6):2272–2286. DOI: 10.1108/ec-04-2017-0128. 8

[23] Chakraverty, S. and Perera, S., 2018. Recent advances in applications of computational and fuzzy mathematics. Springer Nature Singapore. DOI: 10.1007/978-981-13-1153-6. 3

[24] Chakraverty, S., Tapaswini, S., and Behera, D., 2016. Fuzzy differential equations and applications for engineers and scientists. CRC Press. DOI: 10.1201/9781315372853. 3

[25] Comba, J. L. D. and Stol, J., 1993. Affine arithmetic and its applications to computer graphics. In *Proc. of VI SIBGRAPI (Brazilian Symposium on Computer Graphics and Image Processing)*, pages 9–18. 3

[26] Crawford, C. R., 1973. Reduction of a band-symmetric generalized eigenvalue problem. *Communications of the ACM*, 16(1):41–44. DOI: 10.1145/361932.361943. 5

[27] Das, S. and Chakraverty, S., 2012. Numerical solution of interval and fuzzy system of linear equations. *Applications and Applied Mathematics*, 7(1):334–356. 5

[28] De Figueiredo, L. H. and Stolfi, J., 2004. Affine arithmetic: Concepts and applications. *Numerical Algorithms*, 37(1–4):147–158. DOI: 10.1023/b:numa.0000049462.70970.b6. 3

[29] Dehghan, M., Hashemi, B., and Ghatee, M., 2006. Computational methods for solving fully fuzzy linear systems. *Applied Mathematics and Computation*, 179(1):328–343. DOI: 10.1016/j.amc.2005.11.124. 5

[30] Dubois, D. J., 1980. *Fuzzy Sets and Systems: Theory and Applications*, 144. Academic Press. 3

[31] Dubois, D. and Prade, H., 1980. *Fuzzy Sets and Systems: Theory and Applications*, 144. Academic Press. 5

[32] Fazeli, S. A. and Rabiei, F., 2016. Solving nonlinear eigenvalue problems using an improved Newton method. *International Journal of Advanced Computer Science and Applications*, 7(9):438–441. 7

[33] Gao, W., Yang, C., and Meza, J. C., 2009. *Solving a Class of Nonlinear Eigenvalue Problems by Newton's Method*, (No. LBNL-2187E). Lawrence Berkeley National Lab. (LBNL), Berkeley, CA. DOI: 10.2172/965775. 7

[34] Gohberg, I., Lancaster, P., and Rodman, L., 1982. *Matrix Polynomials.* Academic Press, New York. DOI: 10.1137/1.9780898719024. 7

[35] Hansen, E. R., 1975. A generalized interval arithmetic. In *International Symposium on Interval Mathematics*, pages 7–18, Springer, Berlin, Heidelberg. DOI: 10.1007/3-540-07170-9_2. 3

[36] Hanss, M., 2005. Applied Fuzzy Arithmetic: An Introduction with Engineering Applications. Springer, 1:100–116. DOI: 10.1007/b138914. 3

[37] Hladík, M., 2012. Enclosures for the solution set of parametric interval linear systems. *International Journal of Applied Mathematics and Computer Science*, 22(3):561–574. DOI: 10.2478/v10006-012-0043-4. 4

[38] Hladík, M., 2013. Bounds on eigenvalues of real and complex interval matrices. *Applied Mathematics and Computation*, 219(10):5584–5591. DOI: 10.1016/j.amc.2012.11.075. 6

[39] Hladík, M., Daney, D., and Tsigaridas, E., 2011. A filtering method for the interval eigenvalue problem. *Applied Mathematics and Computation*, 217(12):5236–5242. DOI: 10.1016/j.amc.2010.09.066. 6

[40] Hladík, M. and Jaulin, L. 2011. An eigenvalue symmetric matrix contractor. *Reliable Computing*, pages 27–37. 6

[41] Horn, R. A. and Johnson, C. R., 1985. *Matrix Analysis.* Cambridge University Press, Cambridge. DOI: 10.1017/cbo9780511810817. 4

[42] Jaulin, L., Kieffer, M., Didrit, O., and Walter, E., 2001. *Applied Interval Analysis: With Examples in Parameter and State Estimation, Robust Control and Robotics*, 1. Springer-Verlag, London. DOI: 10.1007/978-1-4471-0249-6. 3

[43] Karunakar, P. and Chakraverty, S., 2018. Solving fully interval linear systems of equations using tolerable solution criteria. *Soft Computing*, 22(14):4811–4818. DOI: 10.1007/s00500-017-2668-6. 5

[44] Kaufmann, A. and Gupta, M. M., 1988. *Fuzzy Mathematical Models in Engineering and Management Science.* Elsevier Science Inc. 3

[45] Keramati, B., 2009. An approach to the solution of linear system of equations by He's homotopy perturbation method. *Chaos, Solitons and Fractals*, 41(1):152–156. DOI: 10.1016/j.chaos.2007.11.020. 4

[46] Kolev, L. V., 2006. Outer interval solution of the eigenvalue problem under general form parametric dependencies. *Reliable Computing*, 12(2):121–140. DOI: 10.1007/s11155-006-4875-1. 6

[47] Krämer, W., 2006. Generalized intervals and the dependency problem. In *PAMM: Proc. in Applied Mathematics and Mechanics*, 6:683–684, Wiley Online Library. DOI: 10.1002/pamm.200610322. 3

[48] Kressner, D., 2009. A block Newton method for nonlinear eigenvalue problems. *Numerische Mathematik*, 114(2):355–372. DOI: 10.1007/s00211-009-0259-x. 7

[49] Kurseeva, V. Y., Tikhov, S. V., and Valovik, D. V., 2019. Nonlinear multiparameter eigenvalue problems: Linearised and nonlinearised solutions. *Journal of Differential Equations*, 267(4):2357–2384. DOI: 10.1016/j.jde.2019.03.014. 7

[50] Lawrence, P. W., Van Barel, M., and Van Dooren, P., 2016. Backward error analysis of polynomial eigenvalue problems solved by linearization. *SIAM Journal on Matrix Analysis and Applications*, 37(1):123–144. DOI: 10.1002/pamm.201510282. 7

[51] Leng, H., 2014. Real eigenvalue bounds of standard and generalized real interval eigenvalue problems. *Applied Mathematics and Computation*, 232:164–171. DOI: 10.1016/j.amc.2014.01.070. 6

[52] Leng, H. and He, Z., 2007. Computing eigenvalue bounds of structures with uncertain-but-non-random parameters by a method based on perturbation theory. *Communications in Numerical Methods in Engineering*, 23(11):973–982. DOI: 10.1002/cnm.936. 6

[53] Leng, H. and He, Z., 2010. Computation of bounds for eigenvalues of structures with interval parameters. *Applied Mathematics and Computation*, 216(9):2734–2739. DOI: 10.1016/j.amc.2010.03.121. 6

[54] Leng, H., He, Z., and Yuan, Q., 2008. Computing bounds to real eigenvalues of real-interval matrices. *International Journal for Numerical Methods in Engineering*, 74(4):523–530. DOI: 10.1002/nme.2179. 6

[55] Mahato, N. R. and Chakraverty, S., 2016a. Filtering algorithm for real eigenvalue bounds of interval and fuzzy generalized eigenvalue problems. *ASCE-ASME Journal of Risk and Uncertainty in Engineering Systems, Part B: Mechanical Engineering*, 2(4):044502. DOI: 10.1115/1.4032958. 6

[56] Mahato, N. R. and Chakraverty, S., 2016b. Filtering algorithm for eigenvalue bounds of fuzzy symmetric matrices. *Engineering Computations*, 33(3):855–875. DOI: 10.1108/ec-12-2014-0255. 6

[57] McWilliam, S., 2001. Anti-optimisation of uncertain structures using interval analysis. *Computers and Structures*, 79(4):421–430. DOI: 10.1016/s0045-7949(00)00143-7. 4

[58] Mehrmann, V. and Watkins, D., 2002. Polynomial eigenvalue problems with Hamiltonian structure. *Electronic Transactions on Numerical Analysis*, 13:106–118. 7

[59] Meyer, C. D., 2000. *Matrix Analysis and Applied Linear Algebra*, 71. SIAM. DOI: 10.1137/1.9780898719512. 4

[60] Miyajima, S. and Kashiwagi, M., 2004. A dividing method utilizing the best multiplication in affine arithmetic. *IEICE Electronics Express*, 1(7):176–181. DOI: 10.1587/elex.1.176. 3

[61] Moore, R. E., 1962. Interval arithmetic and automatic error analysis in digital computing. Ph.D. Dissertation, Department of Mathematics, Stanford University. 3

[62] Moore, R. E., 1979. *Methods and Applications of Interval Analysis*, 2. SIAM. DOI: 10.1137/1.9781611970906. 3

[63] Moore, R. E., Kearfott, R. B., and Cloud, M. J., 2009. *Introduction to Interval Analysis*. SIAM Publications, Philadelphia, PA. DOI: 10.1137/1.9780898717716. 3

[64] Neumaier, A., 1986. Linear interval equations. In *Interval Mathematics 1985*. pages 109–120, Springer. DOI: 10.1007/3-540-16437-5_11. 4

[65] Neumaier, A., 1990. *Interval Methods for Systems of Equations*. Cambridge University Press, Cambridge. DOI: 10.1017/cbo9780511526473. 4

[66] Ning, S. and Kearfott, R. B., 1997. A comparison of some methods for solving linear interval equations. *SIAM Journal on Numerical Analysis*, 34(4):1289–1305. DOI: 10.1137/s0036142994270995. 4

[67] Noor, M. A., Noor, K. I., Khan, S., and Waseem, M., 2013. Modified homotopy perturbation method for solving system of linear equations. *Journal of the Association of Arab Universities for Basic and Applied Sciences*, 13(1):35–37. DOI: 10.1016/j.jaubas.2012.07.004. 4

[68] Parlett, B. N., 1998. *The Symmetric Eigenvalue Problem*, 20. SIAM. DOI: 10.1137/1.9781611971163.

[69] Przemieniecki, J. S., 1985. *Theory of Matrix Structural Analysis*. Courier Corporation. 7

[70] Qiu, Z., Chen, S., and Elishakoff, I., 1996. Bounds of eigenvalues for structures with an interval description of uncertain-but-non-random parameters. *Chaos, Solitons and Fractals*, 7(3):425–434. DOI: 10.1016/0960-0779(95)00065-8. 6

[71] Qiu, Z., Chen, S., and Jia, H., 1995. The Rayleigh quotient iteration method for computing eigenvalue bounds of structures with bounded uncertain parameters. *Computers and Structures*, 55:221–227. DOI: 10.1016/0045-7949(94)00444-8. 6

[72] Qiu, Z. and Elishakoff, I., 1998. Antioptimization of structures with large uncertain-but-non-random parameters via interval analysis. *Computer Methods in Applied Mechanics and Engineering*, 152(3–4):361–372. DOI: 10.1016/s0045-7825(96)01211-x. 4

[73] Qiu, Z., Müller, P. C., and Frommer, A., 2001. An approximate method for the standard interval eigenvalue problem of real non-symmetric interval matrices. *Communications in Numerical Methods in Engineering*, 17(4):239–251. DOI: 10.1002/cnm.401. 6

[74] Qiu, Z. and Wang, X., 2005. Solution theorems for the standard eigenvalue problem of structures with uncertain-but-bounded parameters. *Journal of Sound and Vibration*, 282(1–2):381–399. DOI: 10.1016/j.jsv.2004.02.024. 6

[75] Qiu, Z., Wang, X., and Friswell, M. I., 2005. Eigenvalue bounds of structures with uncertain-but-bounded parameters. *Journal of Sound and Vibration*, 282:297–312. DOI: 10.1016/j.jsv.2004.02.051. 6

[76] Rahgooy, T., Sadoghi Yazdi, H., and Monsefi, R., 2009. Fuzzy complex system of linear equations applied to circuit analysis. *International Journal of Computer and Electrical Engineering*, 1(5):535. DOI: 10.7763/ijcee.2009.v1.82. 5

[77] Rajakumar, C., 1993. Lanczos algorithm for the quadratic eigenvalue problem in engineering applications. *Computer Methods in Applied Mechanics and Engineering*, 105(1):1–22. DOI: 10.1016/0045-7825(93)90113-c. 7

[78] Rajakumar, C. and Rogers, C. R., 1991. The Lanczos algorithm applied to unsymmetric generalized eigenvalue problem. *International Journal for Numerical Methods in Engineering*, 32(5):1009–1026. DOI: 10.1002/nme.1620320506. 6

[79] Rex, G. and Rohn, J., 1998. Sufficient conditions for regularity and singularity of interval matrices. *SIAM Journal on Matrix Analysis and Applications*, 20(2):437–445. DOI: 10.1137/s0895479896310743. 6

[80] Rohn, J., 1998. Bounds on eigenvalues of interval matrices. *ZAMM-Zeitschrift fur Angewandte Mathematik und Mechanik*, 78(3):S1049. DOI: 10.1002/zamm.19980781593. 6

[81] Romero-Quete, D. and Cañizares, C. A., 2018. An affine arithmetic-based energy management system for isolated microgrids. *IEEE Transactions on Smart Grid*, 10(3):2989–2998. DOI: 10.1109/tsg.2018.2816403. 4

[82] Rout, S. and Chakraverty, S., 2019. Solving fully fuzzy nonlinear eigenvalue problems of damped spring-mass structural systems using novel fuzzy-affine approach. *Computer Modeling in Engineering and Sciences*, 121(3):947–980. DOI: 10.32604/cmes.2019.08036. 8

[83] Rout, S. and Chakraverty, S., 2020. Affine approach to solve nonlinear eigenvalue problems of structures with uncertain parameters. In *Recent Trends in Wave Mechanics and Vibrations*, pages 407–425, Springer, Singapore. DOI: 10.1007/978-981-15-0287-3_29. 8

[84] Rump, S. M. and Kashiwagi, M., 2015. Implementation and improvements of affine arithmetic. *Nonlinear Theory and its Applications, IEICE*, 6(3):341–359. DOI: 10.1587/nolta.6.341. 4

[85] Saad, Y., El-Guide, M., and Miedlar, A., 2019. A rational approximation method for the nonlinear eigenvalue problem. *ArXiv Preprint ArXiv:1901.01188*. 7

[86] Sadangi, S., 2013. Interval nonlinear eigenvalue problems. M.Sc. thesis, National Institute of Technology Rourkela, India. `http://ethesis.nitrkl.ac.in/5161/` 8

[87] Sim, J., Qiu, Z., and Wang, X., 2007. Modal analysis of structures with uncertain but bounded parameters via interval analysis. *Journal of Sound and Vibration*, 303:29–45. DOI: 10.1016/j.jsv.2006.11.038. 6

[88] Skalna, I., 2009. Direct method for solving parametric interval linear systems with non-affine dependencies. In *International Conference on Parallel Processing and Applied Mathematics*, pages 485–494, Springer, Berlin, Heidelberg. DOI: 10.1007/978-3-642-14403-5_51. 4

[89] Skalna, I. and Hladík, M., 2017. A new algorithm for Chebyshev minimum-error multiplication of reduced affine forms. *Numerical Algorithms*, 76(4):1131–1152. DOI: 10.1007/s11075-017-0300-6. 4

[90] Solovév, S. I. and Solovév, P. S., 2018. Finite element approximation of the minimal eigenvalue of a nonlinear eigenvalue problem. *Lobachevskii Journal of Mathematics*, 39(7):949–956. DOI: 10.1134/s199508021807020x. 7

[91] Stewart, G. W., 1975. Gershgorin theory for the generalized eigenvalue problem $Ax = \lambda Bx$. *Mathematics of Computation*, pages 600–606. DOI: 10.2307/2005580. 5

[92] Stolfi, J. and De Figueiredo, L. H., 2003. An introduction to affine arithmetic. *Trends in Applied and Computational Mathematics*, 4(3):297–312. DOI: 10.5540/tema.2003.04.03.0297. 3

[93] Strang, G., 1993. *Introduction to Linear Algebra*, 3rd ed. Wellesley-Cambridge Press Wellesley, MA. 4

[94] Su, Y. and Bai, Z., 2011. Solving rational eigenvalue problems via linearization. *SIAM Journal on Matrix Analysis and Applications*, 32(1):201–216. DOI: 10.1137/090777542. 7

[95] Tanabe, K., 1971. Projection method for solving a singular system of linear equations and its applications. *Numerische Mathematik*, 17(3):203–214. DOI: 10.1007/bf01436376. 4

[96] Tisseur, F., 2000. Backward error and condition of polynomial eigenvalue problems. *Linear Algebra and its Applications*, 309(1–3):339–361. DOI: 10.1016/s0024-3795(99)00063-4. 7

[97] Tisseur, F. and Meerbergen, K., 2001. The quadratic eigenvalue problem. *SIAM Review*, 43(2):235–286. DOI: 10.1137/s0036144500381988. 7

[98] Wang, S., Wang, K., Teng, F., Strbac, G., and Wu, L., 2018. An affine arithmetic-based multi-objective optimization method for energy storage systems operating in active distribution networks with uncertainties. *Applied Energy*, 223:215–228. DOI: 10.1016/j.apenergy.2018.04.037. 4

[99] Ward, R. C., 1981. Balancing the generalized eigenvalue problem. *SIAM Journal on Scientific and Statistical Computing*, 2(2):141–152. DOI: 10.1137/0902012. 5

[100] Wetherhold, R. and Padliya, P. S., 2014. Design aspects of nonlinear vibration analysis of rectangular orthotropic membranes. *Journal of Vibration and Acoustics*, 136(3):034506. DOI: 10.1115/1.4027148. 7

[101] Wilkinson, J. H., 1965. *The Algebraic Eigenvalue Problem*, 662. Clarendon, Oxford. DOI: 10.2307/2007453. 5

[102] Xia, Y. and Friswell, M., 2014. Efficient solution of the fuzzy eigenvalue problem in structural dynamics. *Engineering Computations*, 31(5):864–878. DOI: 10.1108/ec-02-2013-0052. 6

[103] Xu, C., Gu, W., Gao, F., Song, X., Meng, X., and Fan, M., 2016. Improved affine arithmetic based optimisation model for interval power flow analysis. *IET Generation, Transmission and Distribution*, 10(15):3910–3918. DOI: 10.1049/iet-gtd.2016.0601. 4

[104] Yusufoğlu, E., 2009. An improvement to homotopy perturbation method for solving system of linear equations. *Computers and Mathematics with Applications*, 58(11–12):2231-2235. DOI: 10.1016/j.camwa.2009.03.010. 4

[105] Zadeh, L. A., 1965. Fuzzy sets. *Information and Control*, 8(3):338–353. DOI: 10.1016/s0019-9958(65)90241-x. 3

[106] Zadeh, L. A., Fu, K. S., and Tanaka, K. (Eds.), 2014. Fuzzy sets and their applications to cognitive and decision processes. *Proc. of the U.S.–Japan Seminar on Fuzzy Sets and their Applications*, Academic Press, University of California, Berkeley, CA, July 1–4, 1974. DOI: 10.1016/c2013-0-11734-5. 3

[107] Zimmermann, H. J., 2011. *Fuzzy Set Theory and its Applications*. Springer Science and Business Media. DOI: 10.1007/978-94-015-8702-0. 3

CHAPTER 2

Preliminaries

In this chapter, basic definitions, notations, and terminologies of interval analysis and fuzzy set theory have been incorporated. This chapter starts with interval and some of its terminologies. Then, the interval set operations and their arithmetic with algebraic properties with various examples are introduced. Furthermore, fuzzy set with its terminologies, fuzzy set operations, fuzzy arithmetic, and different types of fuzzy numbers are discussed.

2.1 INTERVAL

The floating-point representation of a real quantity "x" may be referred to as an interval and is denoted by $[x] = [x_{\min}, x_{\max}]$ or $[x] = [\underline{x}, \overline{x}]$, where $\underline{x}$ and $\overline{x}$ are called as the respective lower and upper bounds of the interval $[x]$. From Fig. 2.1, it may be observed that the intervals are subsets of the real number set $\mathbb{R}$. Thus, an interval $[x]$ is a set which may be defined as

$$[x] = [\underline{x}, \overline{x}] = \{\alpha \in \mathbb{R} | \underline{x} \leq \alpha \leq \overline{x}\}, \tag{2.1}$$

where $\underline{x}, \overline{x} \in \mathbb{R}$ and $\underline{x} \leq \overline{x}$.

In general, there exist various types of intervals:

1. **Open interval** ($(x) = (\underline{x}, \overline{x})$), in which its endpoints are not included and is written with parentheses.
2. **Closed interval** ($[x] = [\underline{x}, \overline{x}]$), in which all its limit points are included (given in Eq. (2.1)).
3. **Half-open interval** ($(x] = (\underline{x}, \overline{x}]$ or $[x) = [\underline{x}, \overline{x})$), in which only one of its endpoints is included and is written by mixing the notations for open and closed intervals.

Our work is particularly based upon closed intervals. Thus, throughout the book, we refer to "closed interval" as "interval."

2.2 BASIC TERMINOLOGIES OF INTERVAL

In this section, basic terminologies associated with intervals are included.

2.2.1 INTERVAL CENTER (x_c)

The mid-point of the interval $[x] = [\underline{x}, \overline{x}]$ is called its interval center and may be defined as

$$x_c = \frac{1}{2}(\underline{x} + \overline{x}). \tag{2.2}$$

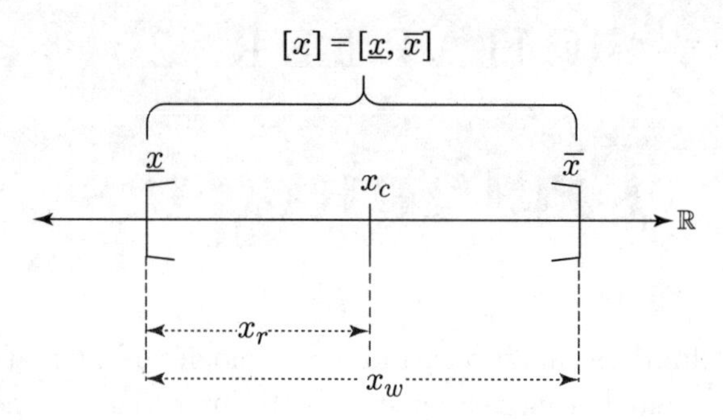

Figure 2.1: Interval $[x] = [\underline{x}, \overline{x}]$.

2.2.2 INTERVAL WIDTH (x_w)

The length between the lower bound and the upper bound of the interval $[x] = [\underline{x}, \overline{x}]$ is referred to as its interval width and may be defined as

$$x_w = \overline{x} - \underline{x}. \tag{2.3}$$

2.2.3 INTERVAL RADIUS (x_Δ)

The half-length of the width of an interval $[x] = [\underline{x}, \overline{x}]$ (Fig. 2.1) is called its interval radius and may be defined as

$$x_\Delta = \frac{1}{2}\left(\overline{x} - \underline{x}\right). \tag{2.4}$$

Note 2.1

The interval $[x] = [\underline{x}, \overline{x}]$ may also be represented in the form of its interval center (x_c) and interval radius (x_Δ) as follows:

$$[x] = [x_c - x_\Delta, x_c + x_\Delta]. \tag{2.5}$$

Example 2.2 Find the interval center, width, and radius of the interval $[x] = [15, 32]$.

Solution: Here, the respective lower and upper bounds of the given interval are $\underline{x} = 15$ and $\overline{x} = 32$. Thus, the interval center, width, and radius are found as follows:

Interval center: $x_c = \frac{1}{2}\left(\underline{x} + \overline{x}\right) = \frac{1}{2}(15 + 32) = 23.5$.

Interval width: $x_w = \overline{x} - \underline{x} = 32 - 15 = 17$.

Interval radius: $x_\Delta = \frac{1}{2}\left(\overline{x} - \underline{x}\right) = \frac{1}{2}(32 - 17) = 8.5$.

2.3 EQUALITY OF INTERVALS

Let us consider two intervals viz. $[x] = [\underline{x}, \overline{x}]$ and $[y] = [\underline{y}, \overline{y}]$. These two intervals are said to be equal if and only if the respective lower and upper bounds are equal to each other. That is,

$$\underline{x} = \underline{y} \quad \text{and} \quad \overline{x} = \overline{y}. \tag{2.6}$$

Moreover, if the lower and upper bounds of an interval $[x] = [\underline{x}, \overline{x}]$ are equal (that is $\underline{x} = \overline{x}$), then it is known as a degenerate interval and may be denoted as $\{x\}$. Hence, the degenerate intervals are identical to the real numbers.

2.4 INTERVAL SET OPERATIONS

In this section, several set operations of intervals viz. interval union, intersection, and hull are addressed. Let us consider two intervals $[x] = [\underline{x}, \overline{x}]$ and $[y] = [\underline{y}, \overline{y}]$. Then, the followings are different interval set operations.

2.4.1 INTERVAL UNION

The union of two intervals $[x]$ and $[y]$ (if they have some common points) may be defined as (given in Fig. 2.2)

$$\begin{aligned} [x] \cup [y] &= \{\alpha | \alpha \in [x] \quad \text{or} \quad \alpha \in [y]\} \\ &= \left[\min\{\underline{x}, \underline{y}\}, \max\{\overline{x}, \overline{y}\}\right]. \end{aligned} \tag{2.7}$$

2.4.2 INTERVAL INTERSECTION

The intersection of two intervals $[x]$ and $[y]$ (if they are having some common points) may be defined as (given in Fig. 2.2)

$$\begin{aligned} [x] \cap [y] &= \{\alpha | \alpha \in [x] \quad \text{and} \quad \alpha \in [y]\} \\ &= \left[\max\{\underline{x}, \underline{y}\}, \min\{\overline{x}, \overline{y}\}\right]. \end{aligned} \tag{2.8}$$

Note 2.3

If there are no common points between the two intervals $[x]$ and $[y]$ (that is $\overline{x} < \underline{y}$ or $\overline{y} < \underline{x}$), then the intersection of these intervals will be empty, that is,

$$[x] \cap [y] = \phi. \tag{2.9}$$

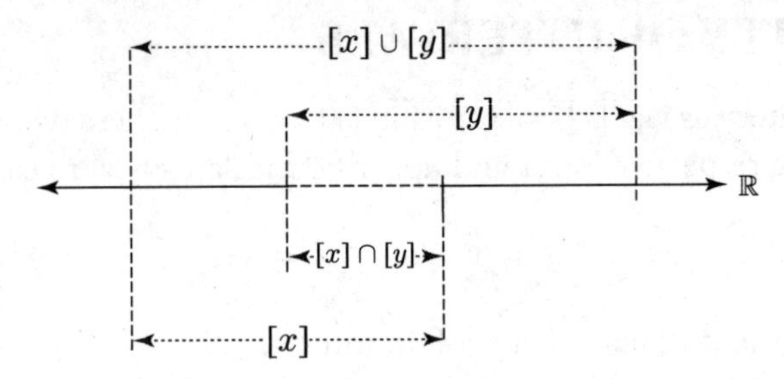

Figure 2.2: The interval union and the interval intersection of $[x]$ and $[y]$.

Example 2.4 Compute the interval union and interval intersection of the intervals $[x] = [3, 19]$ and $[y] = [13, 47]$.

Solution: Here, the lower and upper bounds of the given two intervals $[x]$ and $[y]$ are $\underline{x} = 3$; $\overline{x} = 19$; and $\underline{y} = 13$; $\overline{y} = 47$, respectively. Then, the interval union (from Eq. (2.7)) is

$$[x] \cup [y] = \left[\min\{\underline{x}, \underline{y}\}, \max\{\overline{x}, \overline{y}\}\right] = [\min\{3, 13\}, \max\{19, 47\}] = [3, 47].$$

Similarly, the interval intersection (from Eq. (2.8)) is

$$[x] \cap [y] = \left[\max\{\underline{x}, \underline{y}\}, \min\{\overline{x}, \overline{y}\}\right] = [\max\{3, 13\}, \min\{19, 47\}] = [13, 19].$$

2.4.3 INTERVAL HULL

The interval hull of two intervals $[x]$ and $[y]$ is written as

$$[x]\underline{\cup}[y] = \left[\max\{\underline{x}, \underline{y}\}, \min\{\overline{x}, \overline{y}\}\right]. \tag{2.10}$$

Note 2.5

$$[x]\underline{\cup}[y] \supseteq [x] \cup [y]. \tag{2.11}$$

2.5 INTERVAL ARITHMETIC

In standard interval arithmetic, all the binary operations viz. interval addition, subtraction, scalar multiplications, multiplication, reciprocal, division, and power of an interval are determined in such a way that each calculated interval is guaranteed to all contain the corresponding ideal quantity. That means, when two ideal quantities x and y are contained in their respective intervals $[x]$ and $[y]$, then after any binary operation, the ideal quantity $x * y$ will lie in the interval $[x] * [y]$.

Therefore, if $[x] = [\underline{x}, \overline{x}]$ and $[y] = [\underline{y}, \overline{y}]$ are two interval representations, then all the standard interval arithmetic may be defined as

$$[x] * [y] = \{x * y | x \in [x], y \in [y]\}, \tag{2.12}$$

where "$* = \{+, -, \times, /\}$." Keeping this in view, all the binary operations of the standard interval arithmetic are defined below.

2.5.1 INTERVAL ADDITION

$$[x] + [y] = \left[\underline{x} + \underline{y}, \overline{x} + \overline{y}\right]. \tag{2.13}$$

Example 2.6 Let us consider two intervals $[x] = [4, 9]$ and $[y] = [-12, 2]$. Evaluate the interval sum of $[x]$ and $[y]$.

Solution: The interval bounds of the two given intervals are

$$\underline{x} = 4; \quad \overline{x} = 9 \quad \text{and} \quad \underline{y} = -12; \quad \overline{y} = 2.$$

Let $[s]$ be the interval sum of these given intervals. Thus, $[s]$ may be evaluated by using Eq. (2.13) as follows:

$$[s] = [x] + [y] = \left[\underline{x} + \underline{y}, \overline{x} + \overline{y}\right] = [4 - 12, 9 + 2] = [-8, 11].$$

2.5.2 INTERVAL SUBTRACTION

$$[x] - [y] = \left[\underline{x} - \overline{y}, \overline{x} - \underline{y}\right]. \tag{2.14}$$

Example 2.7 What is the resulting interval, if an interval $[x] = [2, 5]$ is subtracted from itself?

Solution: According to Eq. (2.14), the subtraction of the given interval $[x] = [2, 5]$ from itself is found as

$$[x] - [x] = [2, 5] - [2, 5] = [\underline{x} - \overline{x}, \overline{x} - \underline{x}] = [2 - 5, 5 - 2] = [-3, 3].$$

Note 2.8

From Example 2.7, it is clear that the subtraction of an interval from itself results in a comparatively wider interval instead of being zero. This is due to the dependency of operands in the operation, which is the main reason behind the "interval dependency problem" (explicitly explained in Section 3.1).

2.5.3 INTERVAL SCALAR MULTIPLICATION

$$k \cdot [x] = \begin{cases} [k \cdot \underline{x}, k \cdot \overline{x}], & k \geq 0 \\ [k \cdot \overline{x}, k \cdot \underline{x}] & k < 0, \end{cases} \quad \text{for} \quad k \in \mathbb{R}. \tag{2.15}$$

Example 2.9 Let us consider an interval linear function having three variables as $[f]([x], [y], [z]) = \alpha[x] + \beta[y] + \gamma[z]$. For the values given below, calculate the functional value of the given function:

$$[x] = [6, 11], \quad [y] = [1, 15] \quad \text{and} \quad [z] = [-3, 3]; \quad \alpha = 5, \quad \beta = -9 \quad \text{and} \quad \gamma = 2.$$

Solution: The functional value for the interval linear function is

$$\begin{aligned} [f]([x], [y], [z]) &= \alpha[x] + \beta[y] + \gamma[z] \\ &= 5 \cdot [6, 11] + (-9) \cdot [1, 15] + 2 \cdot [-3, 3] \\ &= [30, 55] - [9, 135] + [-6, 6] \\ &= [24, 61] - [9, 135] = [-111, 52]. \end{aligned}$$

Therefore, $[f]([x], [y], [z]) = [-111, 52]$.

2.5.4 INTERVAL MULTIPLICATION

$$[x] \cdot [y] = [\min\{\Gamma([x], [y])\}, \max\{\Gamma([x], [y])\}], \tag{2.16a}$$

where

$$\Gamma([x], [y]) = \left\{\underline{x} \cdot \underline{y}, \underline{x} \cdot \overline{y}, \overline{x} \cdot \underline{y}, \overline{x} \cdot \overline{y}\right\}. \tag{2.16b}$$

Example 2.10 Find the interval product of the intervals $[x] = [14, 29]$ and $[y] = [-9, -6]$. Also, verify whether interval multiplication satisfies the classical commutative law or not.

Solution: The intervals are given by

$$[x] = [14, 29] \qquad \text{and} \qquad [y] = [-9, -6].$$

Here, the interval bounds of the two given intervals are

$$\underline{x} = 14; \quad \overline{x} = 29 \quad \text{and} \quad \underline{y} = -9; \quad \overline{y} = -6.$$

Thus,

$$\begin{aligned}\Gamma_1([x],[y]) &= \left\{\underline{x}\cdot\underline{y}, \underline{x}\cdot\overline{y}, \overline{x}\cdot\underline{y}, \overline{x}\cdot\overline{y}\right\} \\ &= \{(14)(-9), (14)(-6), (29)(-9), (29)(-6)\} \\ &= \{-126, -84, -261, -174\}.\end{aligned}$$

Suppose, $[p]$ is the interval product of the given interval. Then, $[p]$ may be evaluated as

$$\begin{aligned}[p] = [x]\cdot[y] &= [\min\{\Gamma_1([x],[y])\}, \max\{\Gamma_1([x],[y])\}] \\ &= [\min\{-126, -84, -261, -174\}, \max\{-126, -84, -261, -174\}] \\ &= [-261, -84].\end{aligned}$$

Now to show that the interval product is commutative, let us compute $[q] = [y]\cdot[x]$. So,

$$\begin{aligned}\Gamma_2([y],[x]) &= \left\{\underline{y}\cdot\underline{x}, \underline{y}\cdot\overline{x}, \overline{y}\cdot\underline{x}, \overline{y}\cdot\overline{x}\right\} \\ &= \{(-9)(14), (-6)(14), (-9)(29), (-6)(29)\} \\ &= \{-126, -84, -261, -174\}.\end{aligned}$$

Then,

$$\begin{aligned}[q] = [y]\cdot[x] &= [\min\{\Gamma_2([y]\cdot[x])\}, \max\{\Gamma_2([y]\cdot[x])\}] \\ &= [\min\{-126, -84, -261, -174\}, \max\{-126, -84, -261, -174\}] \\ &= [-261, -84].\end{aligned}$$

Since $[p] = [q]$, hence the interval product obeys commutative property.

2.5.5 INTERVAL RECIPROCAL

$$\frac{1}{[x]} = \begin{cases}\left[\dfrac{1}{\overline{x}}, \dfrac{1}{\underline{x}}\right], & 0 \notin [x] \\ (-\infty, \infty), & 0 \in [x].\end{cases} \tag{2.17}$$

Example 2.11 Calculate the reciprocal of the interval $[x] = [-24, -6]$.

Solution: Here, the bounds of the given interval are $\underline{x} = -24$ and $\overline{x} = -6$. Hence, the reciprocal of $[x]$ is (here, $0 \notin [-24, -6]$)

$$\frac{1}{[x]} = \frac{1}{[-24, -6]} = \left[\frac{1}{\overline{x}}, \frac{1}{\underline{x}}\right] = \left[\frac{-1}{6}, \frac{-1}{24}\right] = [-0.166667, -0.041667].$$

2.5.6 INTERVAL DIVISION

$$\frac{[x]}{[y]} = [x] \cdot \frac{1}{[y]} = \begin{cases} [\underline{x}, \overline{x}] \cdot \left[\frac{1}{\overline{y}}, \frac{1}{\underline{y}}\right] & 0 \notin [y] \\ (-\infty, \infty) & 0 \in [y]. \end{cases} \tag{2.18}$$

Example 2.12 Divide an interval $[x] = [20, 72]$ by itself and evaluate the resulting interval.

Solution: The interval is given by $[x] = [20, 72]$. That is, $\underline{x} = 20$ and $\overline{x} = 72$. Here, we can notice that $0 \notin [20, 72]$. Hence, according to Eq. (2.18), we may have

$$\frac{[x]}{[x]} = [20, 72] \cdot \frac{1}{[20, 72]} = \left[\frac{\underline{x}}{\overline{y}}, \frac{\overline{x}}{\underline{y}}\right] = \left[\frac{20}{72}, \frac{72}{20}\right] = \left[\frac{5}{18}, \frac{18}{5}\right] = [0.277778, 3.6].$$

Note 2.13

Similar to subtraction, the division of an interval by itself results in a wider interval instead of resulting in the degenerated interval $\{1\}$, which may cause an interval dependency problem.

2.5.7 INTERVAL POWER

- If $m > 0$ is an odd integer, then

$$[x]^m = \left[\underline{x}^m, \overline{x}^m\right]. \tag{2.19a}$$

- If $m > 0$ is an even integer, then

$$[x]^m = \begin{cases} \left[\underline{x}^m, \overline{x}^m\right], & [x] > 0 \\ \left[\overline{x}^m, \underline{x}^m\right], & [x] < 0 \\ \left[0, \max\left\{\underline{x}^m, \overline{x}^m\right\}\right], & 0 \in [x]. \end{cases} \tag{2.19b}$$

Example 2.14 Compute the interval cube of the interval $[x] = [-17, 30]$ and the interval biquadrate of the interval $[y] = [-55, 25]$.

Solution: For interval cube, $m = 3 > 0$ and is an odd integer. Thus, the interval cube of $[x]$ is

$$[x]^3 = \left[\underline{x}^3, \overline{x^3}\right] = \left[(-17)^3, (30)^3\right] = [-4913, 27000].$$

Similarly, for interval biquadrate, $m = 4 > 0$ and is an even integer. Therefore, the interval biquadrate of $[y]$ is (here $0 \in [y] = [-55, 25]$)

$$\begin{aligned}[y]^4 = \left[0, \max\left\{\underline{x}^4, \overline{x^4}\right\}\right] &= \left[0, \max\left\{(-55)^4, (25)^4\right\}\right] \\ &= [0, \max\{9150625, 390625\}] = [0, 9150625].\end{aligned}$$

Example 2.15 Find the functional value for the interval nonlinear function (having two variables) $[f]([x], [y]) = [x]^2 + [y]^2 - 7[x][y]$, where $[x] = [-2, -1]$ and $[y] = [4, 8]$ by using different operations of interval arithmetic.

Solution: The given interval nonlinear function is $[f]([x], [y]) = [x]^2 + [y]^2 - 7[x][y]$. Here the values of the variables are $[x] = [-2, -1]$ and $[y] = [4, 8]$. We have to find the functional value of the interval function $[f]([x], [y])$ by substituting the given values of the variables. Thus, the functional value is found as

$$\begin{aligned}[f]([x], [y]) &= [x]^2 + [y]^2 - 7[x][y] \\ &= [-2, -1]^2 + [4, 8]^2 - 7[-2, -1][4, 8] \\ &= [1, 4] + [16, 64] - 7[-16, -4] \\ &= [17, 68] - [-112, -28] \\ &= [17 - (-28), 68 - (-112)] = [45, 180].\end{aligned}$$

Hence, $[f]([x], [y]) = [45, 180]$.

2.6 INTERVAL ALGEBRAIC PROPERTIES

Suppose $[x]$, $[y]$ and $[z]$ are three interval quantities. Then, the algebraic properties for interval systems are explained as follows.

2.6.1 COMMUTATIVE

Intervals are commutative under addition as well as under multiplication. That is,

$$[x] + [y] = [y] + [x]; \tag{2.20a}$$

$$[x] \cdot [y] = [y] \cdot [x]. \tag{2.20b}$$

Note 2.16

From Example 2.10, it may be observed that intervals obey commutative law under interval multiplication.

2.6.2 ASSOCIATIVE

Intervals are associative under addition as well as under multiplication. That is,

$$([x] + [y]) + [z] = [x] + ([y] + [z]); \tag{2.21a}$$

$$([x] \cdot [y]) \cdot [z] = [x] \cdot ([y] \cdot [z]). \tag{2.21b}$$

2.6.3 IDENTITY

The degenerate interval $\{0\}$ is the additive identity, and the degenerate interval $\{1\}$ is the multiplicative identity of an interval $[x]$. That is,

$$[x] + \{0\} = \{0\} + [x] = [x]; \tag{2.22a}$$

$$[x] \cdot \{1\} = \{1\} \cdot [x] = [x]. \tag{2.22b}$$

2.6.4 INVERSE

Since $[x] + (-[x]) = [x] - [x] = [\underline{x} - \overline{x}, \overline{x} - \underline{x}] \neq \{0\}$, there does not exist any additive inverse of the interval $[x]$. Similarly, because $[x] \cdot \frac{1}{[x]} = \frac{[x]}{[x]} = \left[\frac{\underline{x}}{\overline{x}}, \frac{\overline{x}}{\underline{x}}\right] \neq 1$, thus there does not exist any multiplicative inverses of $[x]$.

Note 2.17

The degenerate intervals have both additive and multiplicative inverses. If $\{x\}$ is a degenerated interval, then "$-\{x\}$" is its additive inverse and "$\frac{1}{\{x\}}$" is its multiplicative inverse.

2.6.5 SUB-DISTRIBUTIVE

The intervals sub-distributive law may be defined as

$$[x] \cdot ([y] + [z]) \subseteq [x] \cdot [y] + [x] \cdot [z]. \tag{2.23}$$

Note 2.18

Intervals do not obey the classical distributive law of real numbers.

2.6.6 CANCELLATION LAW

The cancellation law holds in the case of interval addition, but it does not hold for interval multiplication. That is,

$$[x] + [z] = [y] + [z] \Rightarrow [x] = [y]. \tag{2.24}$$

2.7 FUZZY SETS

Fuzzy set is the set of ordered pairs and may be defined as

$$\tilde{F} = \left\{\left(x, \mu_{\tilde{F}}(x)\right) : x \in X,\ \ \mu_{\tilde{F}}(x) \in [0,1]\right\},$$

where $\mu_{\tilde{F}}(x)$ is known as the membership function of $\tilde{F}$ and X be the universal set.

Example 2.19 Suppose $X = \{s_1, s_2, s_3, s_4, s_5, s_6\}$ is a universal set of students. Let the fuzzy set $\tilde{F}$ resemble the set of smart students, in which "smart" is a fuzzy term. Thus, the fuzzy set associated with the universal set resembling the "smart student" may be written in the form of ordered pairs as

$$\tilde{F} = \{(s_1, 0.5), (s_2, 0.9), (s_3, 0.3), (s_4, 0.6), (s_5, 0.4), (s_6, 1)\}.$$

Here, $\tilde{F}$ indicates the smartness of respective students, such as s_1 is 0.5, and so on.

2.8 BASIC TERMINOLOGIES OF FUZZY SET

In this section, a few basic terminologies of fuzzy set $(\tilde{F})$ viz. boundary, support, core, and height are provided [Chakraverty et al. (2019) [2]].

2.8.1 BOUNDARY $(B(\tilde{F}))$

The boundary of a fuzzy set is a crisp set of all the points $(x \in X)$ of the fuzzy set such that

$$B\left(\tilde{F}\right) = \left\{x : \mu_{\tilde{F}}(x) \geq 0\right\}. \tag{2.25}$$

The boundary of fuzzy sets is divided into two parts viz. lower bound and upper bound.

2.8.2 SUPPORT $(S(\tilde{F}))$

The support of a fuzzy set is a crisp set of all the points $(x \in X)$ of the fuzzy set such that

$$S\left(\tilde{F}\right) = \left\{x : \mu_{\tilde{F}}(x) > 0\right\}. \tag{2.26}$$

Note 2.20

The set in which the support is a single point in X with the membership function $\mu_{\tilde{F}}(x) = 1$ is called as fuzzy singleton set.

2.8.3 CORE $(C(\tilde{F}))$

The core of a fuzzy set is a crisp set of all the points $(x \in X)$ of the fuzzy set such that

$$C\left(\tilde{F}\right) = \left\{x : \mu_{\tilde{F}}(x) = 1\right\}. \tag{2.27}$$

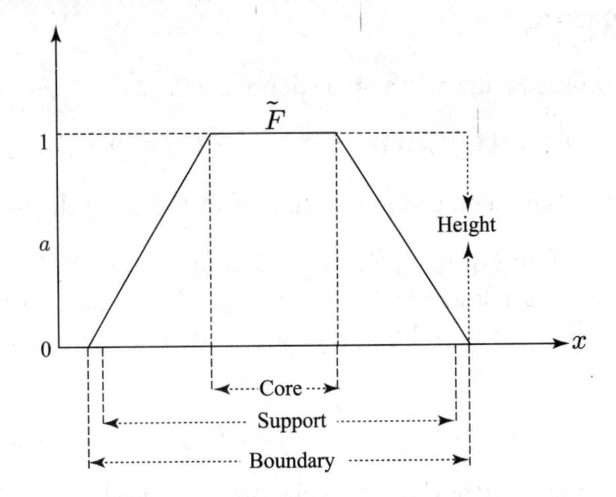

Figure 2.3: Boundary, support, core, and height of the fuzzy set $\tilde{F}$.

2.8.4 HEIGHT ($H(\tilde{F})$)

The height of a fuzzy set is a crisp set of all the points ($x \in X$) of the fuzzy set such that

$$H\left(\tilde{F}\right) = \sup_{x} \left\{\mu_{\tilde{F}}(x)\right\}. \tag{2.28}$$

In Fig. 2.3, the graphical representation of boundary, support, core, and height of the fuzzy set $\left(\tilde{F}\right)$ (given in Section 2.7) is depicted.

2.9 FUZZY SET OPERATIONS

The set operations viz. union, intersection, and complement regarding fuzzy sets are discussed in this section. In this regard, let us consider two fuzzy sets $\tilde{F}_1$ and $\tilde{F}_2$ with the membership functions $\mu_{\tilde{F}_1}(x)$ and $\mu_{\tilde{F}_2}(x)$, respectively.

2.9.1 FUZZY UNION

The fuzzy union of two fuzzy sets $\tilde{F}_1$ and $\tilde{F}_2$ may be defined as

$$\tilde{F}_1 \cup \tilde{F}_2 = \left\{\left(x, \mu_{\tilde{F}_1 \cup \tilde{F}_2}(x)\right) : \mu_{\tilde{F}_1 \cup \tilde{F}_2}(x) = \max\left(\mu_{\tilde{F}_1}(x), \mu_{\tilde{F}_2}(x)\right), \forall x \in X\right\}. \tag{2.29}$$

2.9.2 FUZZY INTERSECTION

The fuzzy intersection of two fuzzy sets $\tilde{F}_1$ and $\tilde{F}_2$ may be defined as

$$\tilde{F}_1 \cap \tilde{F}_2 = \left\{\left(x, \mu_{\tilde{F}_1 \cap \tilde{F}_2}(x)\right) : \mu_{\tilde{F}_1 \cap \tilde{F}_2}(x) = \min\left(\mu_{\tilde{F}_1}(x), \mu_{\tilde{F}_2}(x)\right), \forall x \in X\right\}. \tag{2.30}$$

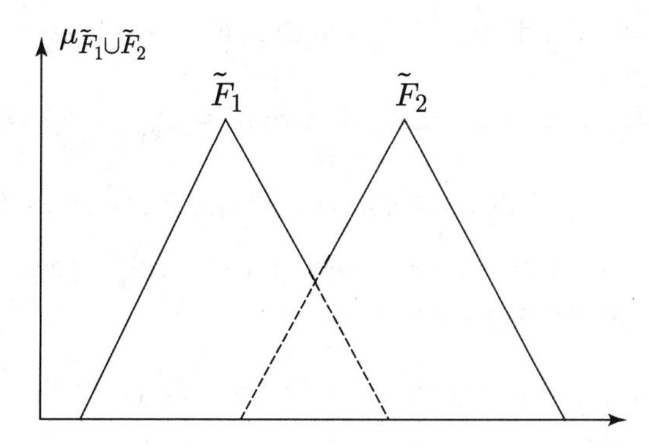

Figure 2.4: The fuzzy union of $\tilde{F}_1$ and $\tilde{F}_2$.

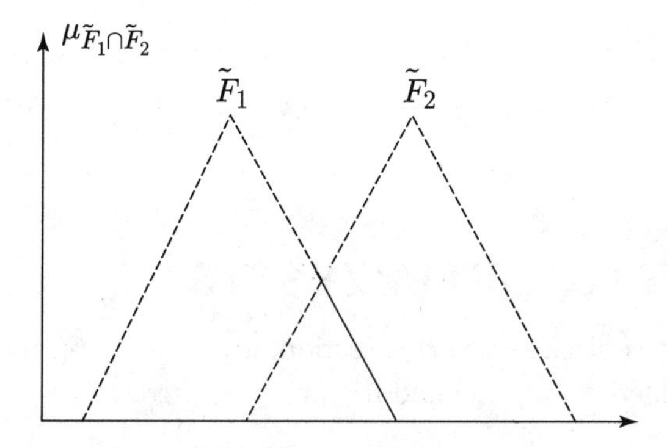

Figure 2.5: The fuzzy intersection of $\tilde{F}_1$ and $\tilde{F}_2$.

2.9.3 FUZZY COMPLEMENT

The fuzzy complement of a fuzzy set $\tilde{F}$ with the membership function $\mu_{\tilde{F}}$ may be denoted as $\tilde{F}^c$ and is expressed as

$$\tilde{F}^c = \left\{\left(x, \mu_{\tilde{F}^c}(x)\right) : \mu_{\tilde{F}^c}(x) = 1 - \mu_{\tilde{F}}(x), \forall x \in X\right\}. \tag{2.31}$$

Example 2.21 Consider two fuzzy sets $\tilde{F}_1$ and $\tilde{F}_2$ with corresponding membership functions $\mu_{\tilde{F}_1}(x)$ and $\mu_{\tilde{F}_2}(x)$ given as follows:

$$\tilde{F}_1 = \{(s_1, 0.5), (s_2, 0.9), (s_3, 0.3), (s_4, 0.6), (s_5, 0.4), (s_6, 1)\} \quad \text{and}$$
$$\tilde{F}_2 = \{(s_1, 0.7), (s_2, 0), (s_3, 1), (s_4, 0.6), (s_5, 0.8), (s_6, 0.5)\}.$$

Compute the fuzzy union, intersection, and complement corresponding to the given fuzzy sets $\tilde{F}_1$ and $\tilde{F}_2$.

Solution: The fuzzy union and intersection of the given two fuzzy sets $\tilde{F}_1$ and $\tilde{F}_2$ are computed as follows.

For the fuzzy union, the membership function is $\mu_{\tilde{F}_1 \cup \tilde{F}_2}(x) = \max(\mu_{\tilde{F}_1}(x), \mu_{\tilde{F}_2}(x))$. Thus,

$$\tilde{F}_1 \cup \tilde{F}_1 = \{(s_1, 0.7), (s_2, 0.9), (s_3, 1), (s_4, 0.6), (s_5, 0.8), (s_6, 1)\}.$$

Similarly, for the fuzzy intersection, the membership function is $\mu_{\tilde{F}_1 \cap \tilde{F}_2}(x) = \min(\mu_{\tilde{F}_1}(x), \mu_{\tilde{F}_2}(x))$. Then, we may have

$$\tilde{F}_1 \cap \tilde{F}_1 = \{(s_1, 0.5), (s_2, 0), (s_3, 0.3), (s_4, 0.6), (s_5, 0.4), (s_6, 0.5)\}.$$

Moreover, for the fuzzy complement, the membership function is $\mu_{\tilde{F}^c}(x) = 1 - \mu_{\tilde{F}}(x)$. Suppose $\tilde{F_1}^c$ and $\tilde{F_2}^c$ denote the fuzzy complement sets of $\tilde{F}_1$ and $\tilde{F}_2$, respectively. Therefore,

$$\tilde{F}_1^c = \{(s_1, 0.5), (s_2, 0.1), (s_3, 0.7), (s_4, 0.4), (s_5, 0.6), (s_6, 0)\} \quad \text{and}$$
$$\tilde{F}_2^c = \{(s_1, 0.3), (s_2, 1), (s_3, 0), (s_4, 0.4), (s_5, 0.2), (s_6, 0.5)\}.$$

2.10 ARITHMETIC OF FUZZY SETS

The fuzzy set arithmetic is discussed in this section, as, suppose $\tilde{F}_1$ and $\tilde{F}_2$ are two fuzzy sets with the membership functions $\mu_{\tilde{F}_1}(x)$ and $\mu_{\tilde{F}_2}(x)$, respectively.

2.10.1 ALGEBRAIC SUM

The algebraic sum of the two fuzzy sets $\tilde{F}_1$ and $\tilde{F}_2$ may be defined as

$$\begin{aligned}\tilde{F}_1 + \tilde{F}_2 = \Big\{\Big(x, \mu_{\tilde{F}_1+\tilde{F}_2}(x)\Big) : \mu_{\tilde{F}_1+\tilde{F}_2}(x) \\ = \mu_{\tilde{F}_1}(x) + \mu_{\tilde{F}_2}(x) - \mu_{\tilde{F}_1}(x) \cdot \mu_{\tilde{F}_2}(x), \forall x \in X\Big\}.\end{aligned} \tag{2.32}$$

2.10.2 ALGEBRAIC PRODUCT

The algebraic product of the two fuzzy sets $\tilde{F}_1$ and $\tilde{F}_2$ is

$$\tilde{F}_1 \times \tilde{F}_2 = \Big\{\Big(x, \mu_{\tilde{F}_1\times\tilde{F}_2}(x)\Big) : \mu_{\tilde{F}_1\times\tilde{F}_2}(x) = \mu_{\tilde{F}_1}(x) \times \mu_{\tilde{F}_2}(x), \forall x \in X\Big\}. \tag{2.33}$$

2.10.3 BOUNDED SUM

The bounded sum of the two fuzzy sets $\tilde{F}_1$ and $\tilde{F}_2$ may be considered as

$$\tilde{F}_1 \oplus \tilde{F}_2 = \Big\{\Big(x, \mu_{\tilde{F}_1\oplus\tilde{F}_2}(x)\Big) : \mu_{\tilde{F}_1\oplus\tilde{F}_2}(x) = \min\Big(1, \mu_{\tilde{F}_1}(x) + \mu_{\tilde{F}_2}(x)\Big), \forall x \in X\Big\}. \tag{2.34}$$

2.10.4 BOUNDED DIFFERENCE

The bounded difference of the two fuzzy sets $\tilde{F}_1$ and $\tilde{F}_2$ may be defined as

$$\begin{aligned}\tilde{F}_1 - \tilde{F}_2 = \Big\{\Big(x, \mu_{\tilde{F}_1 - \tilde{F}_2}(x)\Big) : \mu_{\tilde{F}_1 - \tilde{F}_2}(x) \\ = \max\Big(0, \mu_{\tilde{F}_1}(x) + \mu_{\tilde{F}_2}(x) - 1\Big), \forall x \in X\Big\}.\end{aligned} \tag{2.35}$$

2.10.5 POWER

The mth power (for $m \geq 0$) of a fuzzy set $\tilde{F}$ may be written as

$$\tilde{F}^m = \Big\{(x, \mu_{\tilde{F}^m}(x)) : \mu_{\tilde{F}^m}(x) = (\mu_{\tilde{F}}(x))^m, \forall x \in X\Big\}. \tag{2.36}$$

Example 2.22 Find the algebraic sum and product as well as the bounded sum and difference of the fuzzy sets

$\tilde{F}_1 = \{(1, 0.1), (2, 0.5), (3, 0.8), (4, 1), (5, 0.7), (6, 0.3), (7, 0.1), (8, 0), (9, 0), (10, 0)\}$ and
$\tilde{F}_2 = \{(1, 0), (2, 0.2), (3, 0.8), (4, 1), (5, 1), (6, 0.3), (7, 0), (8, 0), (9, 0), (10, 0)\}$.

Solution: The algebraic sum and product of the given fuzzy set are

$$\tilde{F}_1 + \tilde{F}_2 = \{(1, 0.1), (2, 0.6), (3, 0.96), (4, 1), (5, 1), (6, 0.51), (7, 0.1), (8, 0), (9, 0), (10, 0)\},$$
$$\tilde{F}_1 \times \tilde{F}_2 = \{(1, 0), (2, 0.1), (3, 0.64), (4, 1), (5, 0.7), (6, 0.09), (7, 0), (8, 0), (9, 0), (10, 0)\}.$$

Similarly, the bounded sum and difference of the given fuzzy sets are found as

$$\tilde{F}_1 \oplus \tilde{F}_2 = \{(1, 0.1), (2, 0.7), (3, 1), (4, 1), (5, 1), (6, 0.6), (7, 0.1), (8, 0), (9, 0), (10, 0)\},$$
$$\tilde{F}_1 - \tilde{F}_2 = \{(1, 0), (2, 0), (3, 0.6), (4, 1), (5, 0.7), (6, 0), (7, 0), (8, 0), (9, 0), (10, 0)\}.$$

2.11 FUZZY NUMBERS

The fuzzy number $\tilde{F}$ is defined as a special type of fuzzy set (given Section 2.7) that obeys the following properties:

1. $\tilde{F}$ is normal, (that is $\exists x \in \mathbb{R} : \tilde{F}(x) = 1$);
2. $\tilde{F}$ is convex; and
3. the membership function $\mu_{\tilde{F}}(x)$ is piecewise continuous.

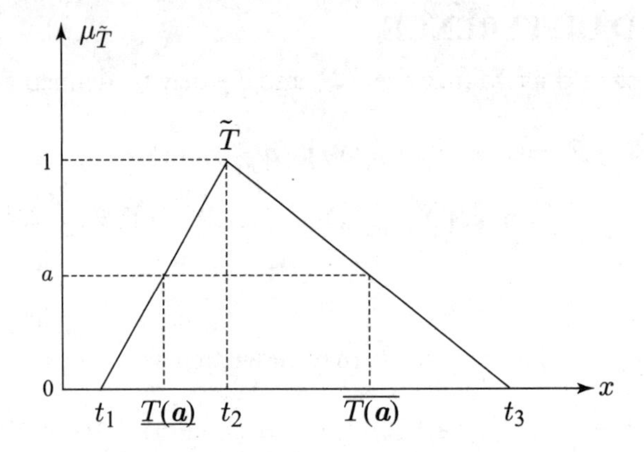

Figure 2.6: Triangular fuzzy number (TFN) $\tilde{T}$.

2.12 DIFFERENT TYPES OF FUZZY NUMBERS

There exist different types of fuzzy numbers viz. triangular fuzzy number (TFN), trapezoidal fuzzy number (TrFN), Gaussian fuzzy number (GFN), and exponential fuzzy number (EFN). In this book, the fuzziness is mainly modeled through TFN and TrFN. As such, the basic definition and behavior of TFN and TrFN are discussed in the following sections.

2.12.1 TRIANGULAR FUZZY NUMBER (TFN)

TFN may be represented by a triplet $\tilde{T} = (t_1, t_2, t_3)$. It generally has a linear graph (that is, the composition of left-increasing and right-decreasing linear functions) as shown in Fig. 2.6. It may be noted that in TFN, there exists exactly one $x_0 \in \mathbb{R}$ such that $\mu_{\tilde{T}}(x_0) = 1$. Here, x_0 is known as the mean value of $\tilde{T}$. The membership function $\mu_{\tilde{T}}(x)$ of a TFN $\tilde{T} = (t_1, t_2, t_3)$ may be defined as follows:

$$\mu_{\tilde{T}}(x) = \begin{cases} 0, & x < t_1,\ x > t_3 \\ \dfrac{x - t_1}{t_2 - t_1}, & x \in [t_1,\ t_2] \\ \dfrac{t_3 - x}{t_3 - t_2}, & x \in [t_2,\ t_3]. \end{cases} \tag{2.37}$$

Note 2.23

1. t_1 and t_3, respectively, are the lower and upper bounds of the TFN $\tilde{T}$.

2. The membership values at the lower and upper bounds of the TFN $\tilde{T}$ are $\mu_{\tilde{T}}(t_1) = 0 = \mu_{\tilde{T}}(t_3)$.

3. The membership values at $x = t_2$ is $\mu_{\tilde{T}}(t_2) = 1$.

4. The TFN $\tilde{T} = (t_1, t_2, t_3)$ will be called a non-negative symmetric TFN if $t_1, t_2, t_3 \in \mathbb{R}^+$ and $t_2 - t_1 = t_3 - t_2$.

Example 2.24 Consider a TFN $\tilde{T} = (-10, -3, 6)$. Compute the membership function associated with the given TFN.

Solution: Here, $t_1 = -10$, $t_2 = -3$, and $t_3 = 6$. Then, we may have $t_2 - t_1 = 7$ and $t_3 - t_2 = 9$. Therefore, the membership function associated with the given TFN $\tilde{T} = (-10, -3, 6)$ is computed (from Eq. (2.37)) as

$$\mu_{\tilde{T}}(x) = \begin{cases} 0, & x < -10, x > 6 \\ \dfrac{x+10}{7}, & x \in [-10, -3] \\ \dfrac{6-x}{9}, & x \in [-3, 6]. \end{cases}$$

2.12.2 TRAPEZOIDAL FUZZY NUMBER (TRFN)

TrFN may be represented by a quadruplet $\tilde{T}r = (r_1, r_2, r_3, r_4)$. As shown in Fig. 2.7, there is an interval $x \in [r_2, r_3]$ such that $\mu_{\tilde{T}r}(x) = 1$. Thus, the membership function $\mu_{\tilde{T}r}(x)$ may be defined as follows:

$$\mu_{\tilde{T}r}(x) = \begin{cases} 0, & x < r_1, x > r_4 \\ \dfrac{x - r_1}{r_2 - r_1}, & x \in [r_1, r_2] \\ 1, & x \in [r_2, r_3] \\ \dfrac{r_4 - x}{r_4 - r_3}, & x \in [r_3, r_4]. \end{cases} \tag{2.38}$$

Example 2.25 Construct the membership function associated with a TrFN $\tilde{T}r = (-11, -4, 0, 7)$.

Solution: Here, $r_1 = -11$, $r_2 = -4$, $r_3 = 0$, and $r_4 = 7$. Then, $r_2 - r_1 = 7$ and $r_4 - r_3 = 7$. Hence, the membership function associated with the given TrFN $\tilde{T}r = (-11, -4, 0, 7)$ may be constructed as follows:

$$\mu_{\tilde{T}r}(x) = \begin{cases} 0, & x < -11, x > 7 \\ \dfrac{x+11}{7}, & x \in [-11, -4] \\ 1, & x \in [-4, 0] \\ \dfrac{7-x}{7}, & x \in [0, 7]. \end{cases}$$

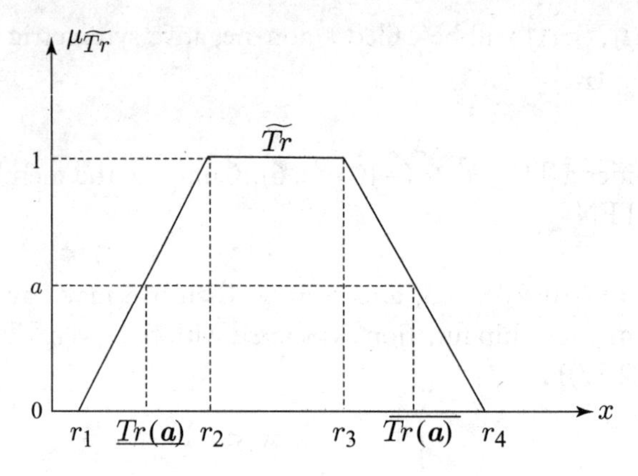

Figure 2.7: Trapezoidal fuzzy number (TrFN) $\tilde{T}r$.

2.13 EXERCISES

2.1. Find the interval $[x]$ if its center (x_c) and width (x_w) are given as $x_c = 0.173926$ and $x_w = 115.235651$.

2.2. Compute the interval sum and product of two intervals $[x] = [59.0012, 96.56]$ and $[y] = [-45.025, -32.755]$. Also, verify whether these intervals obey interval commutative law, associative law, and cancellation law for both addition and multiplication or not.

2.3. For two intervals $[x] = [-20, -10]$ and $[y] = [0, 10]$ find the interval union and interval intersection. Also, compute the interval quantity $[x] \cup ([y] \cap [z])$ for $[z] = [9, 40]$.

2.4. Evaluate the functional value for the interval nonlinear function $[f]([x], [y]) = [x]^2 + 3[y]^2 - 5[z]^2 + [x][y] + 3[z][x] - 13[x][y][z]$, where $[x] = [0.001, 1.99]$, $[y] = [6, 13]$ and $[z] = [-8, -3]$ by using different operations of interval arithmetic.

2.5. Give an example to verify the interval sub-distributive law and interval cancellation law. Further, show that the intervals $[x] = [15, 28]$ and $[y] = [30, 56]$ do not obey interval cancellation law in interval multiplication.

2.6. Evaluate the fuzzy union, intersection, and complement with respect to the fuzzy sets

$$\tilde{F}_1 = \{(1, 0.1), (2, 0.5), (3, 0.8), (4, 1), (5, 0.7), (6, 0.3), (7, 0.1), (8, 0), (9, 0), (10, 0)\}$$

and

$$\tilde{F}_2 = \{(1, 0), (2, 0.2), (3, 0.8), (4, 1), (5, 1), (6, 0.3), (7, 0), (8, 0), (9, 0), (10, 0)\}.$$

2.7. Find the bounded sum and algebraic sum of two fuzzy sets

$$\tilde{F}_1 = \{(s_1, 0.5), (s_2, 0.9), (s_3, 0.3), (s_4, 0.6), (s_5, 0.4), (s_6, 1)\}$$

and

$$\tilde{F}_2 = \{(s_1, 0.7), (s_2, 0), (s_3, 1), (s_4, 0.6), (s_5, 0.8), (s_6, 0.5)\}.$$

2.8. Construct the membership functions associated with the TFN $\tilde{T} = (-23, -15, -8)$ and the TrFN $\tilde{T}r = (1, 9, 16, 22)$.

2.14 REFERENCES

[1] Alefeld, G. and Herzberger, J., 2012. *Introduction to Interval Computations*. Academic Press, London. DOI: 10.1016/C2009-0-21898-8.

[2] Chakraverty, S., Sahoo, D. M., and Mahato, N. R., 2019. Intervals. In *Concepts of Soft Computing*, Springer, Singapore. DOI: 10.1007/978-981-13-7430-2. 29

[3] Hansen, E., 1965. Interval arithmetic in matrix computations, Part I. *Journal of the Society for Industrial and Applied Mathematics, Series B: Numerical Analysis*, 2(2):308–320. DOI: 10.1137/0702025.

[4] Hansen, E. R., 1968. On solving systems of equations using interval arithmetic. *Mathematics of Computation*, 22(102):374–384. DOI: 10.1090/s0025-5718-1968-0229411-4.

[5] Hansen, E. R., 1975. A generalized interval arithmetic. In *International Symposium on Interval Mathematics*, pages 7–18, Springer, Berlin, Heidelberg. DOI: 10.1007/3-540-07170-9_2.

[6] Hanss, M., 2005. *Applied Fuzzy Arithmetic: An Introduction with Engineering Applications*. 1:100–116, Springer. DOI: 10.1007/b138914.

[7] Jaulin, L., Kieffer, M., Didrit, O., and Walter, E., 2001. *Applied Interval Analysis: With Examples in Parameter and State Estimation, Robust Control and Robotics*, 1. Springer-Verlag, London. DOI: 10.1007/978-1-4471-0249-6.

[8] Kaufmann, A. and Gupta, M. M., 1988. *Fuzzy Mathematical Models in Engineering and Management Science*. Elsevier Science Inc.

[9] Krämer, W., 2006. Generalized intervals and the dependency problem. In *PAMM: Proc. in Applied Mathematics and Mechanics*, 6:683–684, Wiley Online Library. DOI: 10.1002/pamm.200610322.

[10] Moore, R. E., 1962. Interval arithmetic and automatic error analysis in digital computing. Ph.D. Dissertation, Department of Mathematics, Stanford University.

[11] Moore, R. E., 1966. *Interval Analysis* (4). Prentice-Hall, Englewood Cliffs, NJ.

[12] Moore, R. E., 1979. *Methods and Applications of Interval Analysis*, 2. SIAM. DOI: 10.1137/1.9781611970906.

[13] Moore, R.E., 1980. Interval methods for nonlinear systems. In *Fundamentals of Numerical Computation (Computer-Oriented Numerical Analysis)*, pages 113–120, Springer, Vienna. DOI: 10.1007/978-3-7091-8577-3_7.

[14] Moore, R. E., Kearfott, R. B., and Cloud, M. J., 2009. *Introduction to Interval Analysis*. SIAM Publications, Philadelphia, PA. DOI: 10.1137/1.9780898717716.

[15] Moore, R. E., Kearfott. R. B., and Cloud, M. J., 2009. *Introduction to Interval Analysis*. SIAM, Philadelphia, PA. DOI: 10.1137/1.9780898717716.

[16] Moore, R. E. and Yang, C. T., 1959. Interval analysis I. *Technical Document LMSD-285875*, Lockheed Missiles and Space Division, Sunnyvale, CA.

[17] Nickel, K. L. (Ed.), 2014. *Interval Mathematics 1980*. Elsevier. DOI: 10.1016/c2013-0-11242-1.

[18] Zadeh, L. A., 1965. Fuzzy sets. *Information and Control*, 8(3):338–353. DOI: 10.1016/s0019-9958(65)90241-x.

[19] Zadeh, L. A., Fu, K. S., and Tanaka, K. (Eds.), 2014. Fuzzy sets and their applications to cognitive and decision processes. *Proc. of the U.S.–Japan Seminar on Fuzzy Sets and their Applications*, Academic Press, University of California, Berkeley, CA, July 1–4, 1974. DOI: 10.1016/c2013-0-11734-5.

[20] Zimmermann, H. J., 2011. *Fuzzy Set Theory—and its Applications*, Springer Science and Business Media. DOI: 10.1007/978-94-015-8702-0.

CHAPTER 3

Affine Arithmetic

In 1993, Comba and Stol gave us the basic idea of affine arithmetic [Comba and Stol (1993) [3]]. The interval dependency problem that occurs in standard interval arithmetic is the main cause behind development of affine arithmetic. Affine arithmetic is a self-validated numerical model that records the range for each ideal quantity and also keeps track of first-order correlations between these quantities. For this additional information, the approximation error is incurred in each operation of affine arithmetic. Therefore, affine arithmetic can overcome the extreme increment of the width of the resulting interval. This benefit will help for several chained-interval computations where interval arithmetic goes through an error explosion. Also, affine arithmetic provides the geometric representation of joint ranges for the related quantities, which may be useful for different interval methods.

3.1 INTERVAL DEPENDENCY PROBLEM

Standard interval arithmetic assumes that all the operands of a computation vary independently over their ranges while performing any interval operations, but the operands may not be independent of each other every time. In these cases when they are partially dependent, standard interval arithmetic overestimates the resulting interval solution into a comparatively wider interval than the exact range. This situation is known as an "interval dependency problem" or "interval overestimation problem." For several complex calculations and long iterative computations, the overestimation range of the solution rapidly increases.

Note 3.1

Since the operation of subtraction in standard interval arithmetic cannot assume that the two operands may denote the same ideal quantity because there may be some cases where two independent operands may have the same range. Thus, the operation of subtraction of standard interval arithmetic may cause an interval dependency problem.

Example 3.2 Let us consider a simple example of the interval dependency problem in standard interval arithmetic. The subtraction of an interval $[x] = [\underline{x}, \overline{x}]$ from itself gives

$$[x] - [x] = [\underline{x} - \overline{x}, \overline{x} - \underline{x}] \neq [0].$$

Here the width of the given interval $[x]$ is $\overline{x} - \underline{x}$ and the width of the resulting interval is $(\overline{x} - \underline{x}) - (\underline{x} - \overline{x}) = 2(\overline{x} - \underline{x})$. Therefore, the width of the resulting interval is twice the width of $[x]$

instead of being zero.

Note 3.3

The interval dependency problem is a major hurdle while handling uncertain real-life problems. In this regard, affine arithmetic has been introduced to handle uncertain parameters efficiently. Affine arithmetic and all its operations and properties are explained in this chapter.

3.2 DEFINITION AND NOTATION

Affine form representation is a linear polynomial of real variables. Suppose the affine form representation of an ideal quantity x may be denoted as $\hat{x}$. Then, $\hat{x}$ is defined as

$$x \in x_0 + \left[-\sum_{i=1}^{n} |x_i|, \sum_{i=1}^{n} |x_i|\right]. \tag{3.1}$$

This may be explicitly written as

$$x \in \hat{x} = x_0 + \sum_{i=1}^{n} x_i \varepsilon_i = x_0 + x_1\varepsilon_1 + \cdots + x_{n-1}\varepsilon_{n-1} + x_n\varepsilon_n, \tag{3.2}$$

where the real variables ε_i for $i = 1, 2, \ldots, n$ are known as noise symbols, and these noise symbols lie in a particular interval $\mathbb{D} = [-1, 1]$. Further, the initial term x_0 is known as the central value of $\hat{x}$ and each associate coefficient x_i for $i = 1, 2, \ldots, n$ of the respective noise symbol ε_i is called the partial deviation of $\hat{x}$.

Moreover, the sum of the magnitude of all the presented partial deviations x_i for $i = 1, 2, \ldots, n$ of the affine form $\hat{x}$ is called the total deviation of $\hat{x}$. The total deviation of $\hat{x}$ may be denoted as d_x and is defined as

$$d_x = \sum_{i=1}^{n} |x_i| = |x_1| + |x_2| + \cdots + |x_n|.$$

Example 3.4 Let $\hat{x} = 57 + 21\varepsilon_1 - \varepsilon_2 - 4\varepsilon_3 + 10\varepsilon_4$ is an affine form representation. Here, 57 is the central value of the affine $\hat{x}$. There exist 4 noise symbols $\varepsilon_i \in \mathbb{D} = [-1, 1]$ for $i = 1, 2, 3, 4$ in the representation $\hat{x}$ and the associated partial deviations are 21, −1, −4, and 10.

Example 3.5 Find the total deviation of the affine form representation $\hat{x}$ given in Example 3.4.

Solution: The total deviation of the affine $\hat{x} = 57 + 21\varepsilon_1 - \varepsilon_2 - 4\varepsilon_3 + 10\varepsilon_4$ is computed as

$$d_x = \sum_{i=1}^{4} |x_i| = |x_1| + |x_2| + |x_3| + |x_4| = 21 + 1 + 4 + 10 = 36.$$

Note 3.6

1. The partial deviations x_i for $i = 1, 2, \ldots, n$ are finite real numbers.
2. All the noise symbols of an affine form representation are unique and independent of each other.
3. In two different affine form representations, there may exist some common noise symbols.
4. The number of noise symbols for different representations may not be equal.
5. New noise symbols may also be generated during affine computations.

3.3 CONVERSION OF AFFINE TO INTERVAL

Let us consider the affine form representation of an ideal quantity x as

$$x \in \hat{x} = x_0 + \sum_{i=1}^{n} x_i \varepsilon_i = x_0 + x_1\varepsilon_1 + \cdots + x_n\varepsilon_n. \tag{3.3}$$

Then the total deviation of $\hat{x}$ is

$$d_x = \sum_{i=1}^{n} |x_i| = |x_1| + |x_2| + \cdots + |x_n|. \tag{3.4}$$

Let $[x] = [\underline{x}, \overline{x}]$ be the interval bound of the affine form representation (3.3). Thus, the lower and upper bounds of $[x]$ may be computed as follows.

- Lower bound of $[x] = \underline{x} = x_0 - d_x$.
- Upper bound of $[x] = \overline{x} = x_0 + d_x$.

Example 3.7 Consider an affine form representation $\hat{x} = 45 - 7\varepsilon_1 + 5\varepsilon_2 + 13\varepsilon_3 - 6\varepsilon_4$. Compute the outer enclosures of the given affine form.

Solution: Here the central value of the given affine form representation $\hat{x}$ is $x_0 = 45$ and the total deviation is obtained as

$$d_x = \sum_{i=1}^{4} |x_i| = 7 + 5 + 13 + 6 = 31.$$

Thus, the lower bound of $\hat{x}$ is found to be

$$\underline{x} = x_0 - d_x = 45 - 31 = 14.$$

and the upper bound of $\hat{x}$ is calculated as

$$\overline{x} = x_0 + d_x = 45 + 31 = 76.$$

Therefore, if $[x]$ is the interval form of the affine representation $\hat{x}$, then $[x] = [14, 76]$.

Example 3.8 Convert the affine form representation $\hat{y} = 24 + 5\varepsilon_1 - 12\varepsilon_3 + \varepsilon_4 - 3\varepsilon_5$ into its interval form.

Solution: $y_0 = 24$ is the central value of $\hat{y}$ and its total deviation is found to be

$$d_y = \sum_{i=1}^{5} |y_i| = 5 + 0 + 12 + 1 + 3 = 21.$$

Suppose $[y]$ be the interval form of $\hat{y}$. Thus, $[y]$ is computed as follows:

$$[y] = [y_0 - d_y, y_0 + d_y] = [24 - 21, 24 + 21] = [3, 45].$$

Note 3.9

1. If $[x]$ is the interval form of the affine representation $\hat{x}$, then $[x]$ is the smallest interval that contains all the possible values of $\hat{x}$ for each noise symbol ranges over the interval $\mathbb{D} = [-1, 1]$ independently.
2. All the correlation information presented in the affine form representation $\hat{x}$ is discarded when it is converted into its interval form $[x]$.

3.4 CONVERSION OF INTERVAL TO AFFINE

Let us consider an interval $[x] = [\underline{x}, \overline{x}]$. If x_0 denotes the center of the interval and x_j is the half-width or radius of the interval, then they may be obtained as

$$x_0 = \frac{1}{2}(\underline{x} + \overline{x}) \quad \text{and} \quad x_j = \frac{1}{2}(\overline{x} - \underline{x}). \tag{3.5}$$

Suppose $\hat{x}$ is the corresponding affine form representation of the given interval $[x]$. Thus, $\hat{x}$ may be expressed as follows:

$$\hat{x} = x_0 + x_j\varepsilon_j, \quad \text{for} \quad \varepsilon_j \in \mathbb{D} = [-1, 1]. \tag{3.6}$$

Note 3.10

1. ε_j is the newly generated noise symbol for the affine representation $\hat{x}$ which lies in the interval $\mathbb{D} = [-1, 1]$.

2. The new noise symbol is used to express the uncertainties present in the ideal quantity of the interval $[x]$.

3. The newly generated noise symbol ε_j should not be present in any other existing affine representations.

Example 3.11 Convert the intervals $[x] = [-4, 5]$ and $[y] = [-2.5, -0.75]$ into their affine form representations.

Solution: The respective affine form representations of the above given intervals are computed as follows:

$$\hat{x} = \frac{1}{2}(-4+5) + \frac{1}{2}(5-(-4))\varepsilon_1 = 0.5 + 4.5\varepsilon_1$$

and

$$\hat{y} = \frac{1}{2}(-2.5+(-0.75)) + \frac{1}{2}(-0.75+2.5)\varepsilon_2 = -1.625 + 0.875\varepsilon_2,$$

where $\varepsilon_1, \varepsilon_2 \in \mathbb{D} = [-1, 1]$ are newly generated noise symbols different from each other.

3.5 AFFINE ARITHMETIC OPERATIONS

This section addresses the binary operations viz. addition, subtraction, multiplication, and division. We have included several example problems to show these operations. More details of the affine arithmetic operations may be found in De Figueiredo and Stolfi (2004) [4] and Skalna (2009) [8]. Readers are encouraged to refer to these works as well.

Suppose $\hat{x}$ and $\hat{y}$ are two affine forms that may be represented as

$$\hat{x} = x_0 + \sum_{i=1}^{n} x_i\varepsilon_i = x_0 + x_1\varepsilon_1 + \cdots + x_n\varepsilon_n, \tag{3.7}$$

and

$$\hat{y} = y_0 + \sum_{i=1}^{n} y_i\varepsilon_i = y_0 + y_1\varepsilon_1 + \cdots + y_n\varepsilon_n. \tag{3.8}$$

Correspondingly, the affine arithmetic operations may be defined as follows:

3.5.1 AFFINE ADDITION

$$\begin{aligned}\hat{x} + \hat{y} &= (x_0 + y_0) + (x_1 + y_1)\,\varepsilon_1 + \cdots + (x_n + y_n)\,\varepsilon_n \\ &= (x_0 + y_0) + \sum_{i=1}^{n} (x_i + y_i)\,\varepsilon_i.\end{aligned} \tag{3.9}$$

3.5.2 AFFINE SUBTRACTION

$$\begin{aligned}\hat{x} - \hat{y} &= (x_0 - y_0) + (x_1 - y_1)\,\varepsilon_1 + \cdots + (x_n - y_n)\varepsilon_n \\ &= (x_0 - y_0) + \sum_{i=1}^{n} (x_i - y_i)\,\varepsilon_i.\end{aligned} \tag{3.10}$$

3.5.3 AFFINE SCALAR MULTIPLICATION

For $a \in \mathbb{R}$,

$$\begin{aligned}a \cdot \hat{x} &= (a \cdot x_0) + (a \cdot x_1)\,\varepsilon_1 + \cdots + (a \cdot x_n)\,\varepsilon_n \\ &= (a \cdot x_0) + \sum_{i=1}^{n} (a \cdot x_i)\,\varepsilon_i.\end{aligned} \tag{3.11}$$

Note 3.12

The affine form representation of the scalar function $f(x, y) = \alpha x + \beta y + \gamma$ for $\alpha, \beta, \gamma \in \mathbb{R}$ may be computed as:

$$\begin{aligned}\hat{f}\,(\hat{x}, \hat{y}) = \alpha\hat{x} + \beta\hat{y} + \gamma &= (\alpha x_0 + \beta y_0 + \gamma) + \sum_{i=1}^{n} (\alpha x_i + \beta y_i)\,\varepsilon_i, \\ &= (\alpha x_0 + \beta y_0 + \gamma) + (\alpha x_1 + \beta y_1)\,\varepsilon_1 + \cdots + (\alpha x_n + \beta y_n)\,\varepsilon_n.\end{aligned} \tag{3.12}$$

Example 3.13 Suppose $\hat{x}$ and $\hat{y}$ are two affine form representations given as $\hat{x} = 45 - 7\varepsilon_1 + 5\varepsilon_2 + 13\varepsilon_3 - 6\varepsilon_4$ and $\hat{y} = 24 + 5\varepsilon_1 - 12\varepsilon_3 + \varepsilon_4 - 3\varepsilon_5$, respectively. Then, perform the affine addition and subtraction for these two affine forms.

Solution: The affine addition operation using Eq. (3.9) is computed as

$$\begin{aligned}\hat{x} + \hat{y} &= (45 - 7\varepsilon_1 + 5\varepsilon_2 + 13\varepsilon_3 - 6\varepsilon_4) + (24 + 5\varepsilon_1 - 12\varepsilon_3 + \varepsilon_4 - 3\varepsilon_5) \\ &= 69 - 2\varepsilon_1 + 5\varepsilon_2 + \varepsilon_3 - 5\varepsilon_4 - 3\varepsilon_5\end{aligned}$$

and the affine subtraction operation using Eq. (3.10) is performed as

$$\begin{aligned}\hat{x} - \hat{y} &= (45 - 7\varepsilon_1 + 5\varepsilon_2 + 13\varepsilon_3 - 6\varepsilon_4) - (24 + 5\varepsilon_1 - 12\varepsilon_3 + \varepsilon_4 - 3\varepsilon_5)\\ &= 21 - 12\varepsilon_1 + 5\varepsilon_2 + 25\varepsilon_3 - 7\varepsilon_4 + 3\varepsilon_5.\end{aligned}$$

Example 3.14 Compute the affine form representation of the linear function $f(x, y) = 6x - 15y + 70$, where the ideal quantities $x \in \hat{x}$ and $y \in \hat{y}$ for the affine representations $\hat{x}$ and $\hat{y}$ are the same as given in Example 3.13.

Solution: The affine linear function representation is performed as

$$\begin{aligned}\hat{f}(\hat{x}, \hat{y}) &= 6\hat{x} - 15\hat{y} + 70\\ &= 6(45 - 7\varepsilon_1 + 5\varepsilon_2 + 13\varepsilon_3 - 6\varepsilon_4) - 15(24 + 5\varepsilon_1 - 12\varepsilon_3 + \varepsilon_4 - 3\varepsilon_5) + 70\\ &= (270 - 360 + 70) + (-42 - 75)\varepsilon_1 + (30 + 0)\varepsilon_2 + (78 + 180)\varepsilon_3\\ &\quad + (-36 - 15)\varepsilon_4 + (0 + 45)\varepsilon_5\\ &= -20 - 117\varepsilon_1 + 30\varepsilon_2 + 258\varepsilon_3 - 51\varepsilon_4 + 45\varepsilon_5.\end{aligned}$$

Example 3.15 Find the interval bounds of $26[a] - 30[b]$ for $[a] = [-8, -3]$ and $[b] = [2, 10]$ by using affine arithmetic computations.

Solution: Converting the given intervals $[a] = [-8, -3]$ and $[b] = [2, 10]$ into their respective affine form representation by using Eq. (3.6), we may have

$$\hat{a} = -5.5 + 2.5\varepsilon_1 \quad \text{and} \quad \hat{b} = 6 + 4\varepsilon_2, \quad \text{for} \quad \varepsilon_1, \varepsilon_2 \in \mathbb{D} = [-1, 1].$$

Suppose $[f] = 26[a] - 30[b]$. Then, the affine form of the function is obtained as

$$\begin{aligned}\hat{f} &= 26\hat{a} - 30\hat{b} = 26(-5.5 + 2.5\varepsilon_1) - 30(6 + 4\varepsilon_2)\\ &= -323 + 65\varepsilon_1 - 120\varepsilon_2.\end{aligned}$$

3.5.4 AFFINE MULTIPLICATION

$$\hat{x} \cdot \hat{y} = x_0 y_0 + \sum_{i=1}^{n} (x_0 y_i + x_i y_0)\,\varepsilon_i + z_j \varepsilon_j \tag{3.13}$$

where

$$|z_j| \geq \left|\sum_{i=1}^{n} x_i \varepsilon_i \cdot \sum_{i=1}^{n} y_i \varepsilon_i\right|, \quad \varepsilon_i \in \mathbb{D} = [-1, 1]. \tag{3.14}$$

Because,

$$\begin{aligned}\hat{x}\cdot\hat{y} &= \left(x_0+\sum_{i=1}^{n}x_i\varepsilon_i\right)\cdot\left(y_0+\sum_{i=1}^{n}y_i\varepsilon_i\right)\\ &= x_0y_0+\sum_{i=1}^{n}(x_0y_i+y_0x_i)\,\varepsilon_i+\sum_{i=1}^{n}x_i\varepsilon_i\cdot\sum_{i=1}^{n}y_i\varepsilon_i.\end{aligned} \tag{3.15}$$

Here, ε_j is the newly generated noise symbol during multiplication and $|z_j|$ is the upper bound of the approximation error.

Note 3.16

There are some other methods for affine multiplication such as Chebyshev minimum-error multiplication developed by Skalna and Hladík (2017) [9]. Chebyshev minimum-error multiplication of affine forms may result in a better approximation than that of the general multiplication of affine arithmetic. Readers may also refer to the work of Skalna and Hladík (2017) [9] for more details about the Chebyshev minimum-error multiplication.

3.5.5 AFFINE DIVISION

$$\frac{\hat{x}}{\hat{y}}=\hat{x}\cdot\frac{1}{\hat{y}}=\frac{x_0}{y_0}+\frac{1}{\hat{y}}\sum_{i=1}^{n}\left(x_i-\frac{x_0}{y_0}y_i\right)\varepsilon_i,\quad\text{provided}\quad \hat{y}\neq\{0\}. \tag{3.16}$$

Example 3.17 Perform the affine multiplication operation of the two affine form representations $\hat{x}=45-7\varepsilon_1+5\varepsilon_2+13\varepsilon_3-6\varepsilon_4$ and $\hat{y}=24+5\varepsilon_1-12\varepsilon_3+\varepsilon_4-3\varepsilon_5$.

Solution: The affine multiplication of the two given affine forms is performed as follows:

$$\begin{aligned}\hat{x}\cdot\hat{y} &= (45-7\varepsilon_1+5\varepsilon_2)\cdot(24+5\varepsilon_1-12\varepsilon_3)\\ &= 1080+225\varepsilon_1-540\varepsilon_3-168\varepsilon_1-35\,(\varepsilon_1)^2+84\varepsilon_1\varepsilon_3+120\varepsilon_2+25\varepsilon_2\varepsilon_1-60\varepsilon_2\varepsilon_3\\ &= 1080+57\varepsilon_1+120\varepsilon_2-540\varepsilon_3-35\varepsilon_4+84\varepsilon_5+25\varepsilon_6-60\varepsilon_7,\end{aligned}$$

where $\varepsilon_i\in[-1,1]$ for $i=1,\ldots,7$. Here, $\varepsilon_4=(\varepsilon_1)^2$, $\varepsilon_5=\varepsilon_1\varepsilon_3$, $\varepsilon_6=\varepsilon_2\varepsilon_1$ and $\varepsilon_7=\varepsilon_2\varepsilon_3$ are newly generated noise symbols during the affine multiplication.

Example 3.18 Find the square of an interval $[a]=[5,10]$ using affine arithmetic operations.

Solution: After conversion, the interval $[a]=[5,10]$ may be obtained as

$$\hat{a}=7.5+2.5\varepsilon_1,\quad\text{for}\quad\varepsilon_1\in[-1,1].$$

Then, the square of the affine form is performed as follows:

$$(\hat{a})^2 = (7.5 + 2.5\varepsilon_1)^2 = 56.25 + 37.5\varepsilon_1 + 6.25\varepsilon_2,$$

where $\varepsilon_2 = (\varepsilon_1)^2 \in [-1, 1]$ is the newly generated noise symbol while operating.

Example 3.19 Using affine arithmetic, evaluate the value of $[x][y] - [y][x]$, where $[x] = [0, 2]$ and $[y] = [1, 3]$.

Solution: The affine form representations of the given intervals are

$$\hat{x} = 1 + \varepsilon_1 \quad \text{and} \quad \hat{y} = 2 + \varepsilon_2, \quad \text{for} \quad \varepsilon_1, \varepsilon_2 \in [-1, 1].$$

Thus, by using affine arithmetic, the value of $[x][y] - [y][x]$ may be evaluated as

$$\begin{aligned} \hat{x}\hat{y} - \hat{y}\hat{x} &= (1 + \varepsilon_1)(2 + \varepsilon_2) - (2 + \varepsilon_2)(1 + \varepsilon_1) \\ &= (2 + 2\varepsilon_1 + \varepsilon_2 + 2\varepsilon_3) - (2 + 2\varepsilon_1 + \varepsilon_2 + 2\varepsilon_3) = 0. \end{aligned}$$

3.6 EFFICACY OF AFFINE ARITHMETIC

Affine arithmetic is an efficient tool to overcome the "interval dependency problem" in the case of standard interval arithmetic. In order to show the efficacy and reliability of affine arithmetic, some examples are included in the present section.

The best way to show the efficiency of affine arithmetic is through the operation of subtraction. As mentioned in Section 3.1, the subtraction of an interval from itself through using standard interval arithmetic results in a comparatively wider interval instead of being zero. But, using affine arithmetic, the subtraction of the interval from itself results in 0.

Example 3.20 Suppose $[x] = [3, 6]$ is an interval. The affine form representation of the interval is $\hat{x} = 4.5 + 1.5\varepsilon_1$. The subtraction of $[x]$ from itself by using standard interval arithmetic gives

$$[x] - [x] = [3, 6] - [3, 6] = [3 - 6, 6 - 3] = [-3, 3],$$

while by using affine computations one may have

$$\hat{x} - \hat{x} = (4.5 + 1.5\varepsilon_1) - (4.5 + 1.5\varepsilon_1) = 0.$$

Hence, affine arithmetic results in tighter bounds.

Example 3.21 Let us consider an interval linear function $[f] = \alpha[x] + \beta[y] + \gamma[z] + \delta$, such that $[x] = [-5, -1]$, $[y] = [1, 2]$ and $[z] = [3, 6]$. Further, all the coefficients of the given linear function are considered in the form of crisp numbers such as $\alpha = 10$, $\beta = -6$, $\gamma = 2$, and $\delta = -7$. Perform both interval and affine arithmetic to compute the functional values of the given interval linear function.

Solution: Performing standard interval arithmetic operations (as mentioned in Chapter 2), the functional value of the given interval linear function $[f]$ is found as

$$\begin{aligned}[f] &= \alpha[x] + \beta[y] + \gamma[z] + \delta \\ &= 10[-5,-1] + (-6)[1,2] + 2[3,6] + (-7) \\ &= [-63,-11]\end{aligned}$$

$\Rightarrow [f] = [-63,-11]$.

Further, to perform affine arithmetic, we have to transfer the given variables into their respective affine forms as follows:

$$\hat{x} = -3 + 2\varepsilon_1,\ \hat{y} = 1.5 + 0.5\varepsilon_2 \quad \text{and} \quad \hat{z} = 4.5 + 1.5\varepsilon_3,$$

where $\varepsilon_i = [-1,1]$, for $i = 1,2,3$ are the respective noise symbols of the above affine forms.

Thus, by adopting affine arithmetic, the functional value of the given interval linear function is computed as

$$\begin{aligned}\hat{f} &= \alpha\hat{x} + \beta\hat{y} + \gamma\hat{z} + \delta \\ &= 10\,(-3 + 2\varepsilon_1) + (-6)(1.5 + 0.5\varepsilon_2) + 2\,(4.5 + 1.5\varepsilon_3) + (-7) \\ &= -37 + 20\varepsilon_1 - 3\varepsilon_2 + 3\varepsilon_3\end{aligned}$$

$\Rightarrow \hat{f} = -37 + 20\varepsilon_1 - 3\varepsilon_2 + 3\varepsilon_3$, where $\varepsilon_i = [-1,1]$, for $i = 1,2,3$.

Converting the above affine solution into its interval bounds, one may have

$$\hat{f} = [-63,-11].$$

Therefore, for the case of linear function the solutions occurred by both interval and affine arithmetic are the same. That is,

$$\hat{f} = [f].$$

Example 3.22 Let us consider a nonlinear function such that $f(x,y) = x^2 + y^2 - xy - y$, where $\forall x \in [0,10]$ and $\forall y \in [3,8]$. Find the functional values of the nonlinear function by using both interval and affine arithmetic. Finally, show that the affine arithmetic is more efficient and may give a tighter enclosure than the interval arithmetic.

Solution: The functional value of the given nonlinear equation is computed by using standard interval arithmetic as follows:

$$\begin{aligned}[f]([x],[y]) &= [x]^2 + [y]^2 - [x][y] - [y] \\ &= [0,10]^2 + [3,8]^2 - [0,10][3,8] - [3,8] \\ &= [-79,161].\end{aligned}$$

Therefore, for standard interval arithmetic, $f(x,y) \in [-79,161]$.

On the other hand, the interval bounds of the given variables are transformed into their respective affine forms (as mentioned in Section 3.4) as

$$[x] = [0,10] \Rightarrow \hat{x} = 5 + 5\varepsilon_1 \quad \text{and} \quad [y] = [3,8] \Rightarrow \hat{y} = 5.5 + 2.5\varepsilon_2, \quad \text{for} \quad \varepsilon_1, \varepsilon_2 \in [-1,1].$$

Now, by adopting the operations of affine arithmetic, the affine functional value of the given nonlinear function $f(x,y)$ is

$$\begin{aligned}\hat{f}(\hat{x},\hat{y}) &= \hat{x}^2 + \hat{y}^2 - \hat{x}\hat{y} - \hat{y} \\ &= (5+5\varepsilon_1)^2 + (5.5+2.5\varepsilon_2)^2 - (5+5\varepsilon_1)(5.5+2.5\varepsilon_2) - (5.5+2.5\varepsilon_2) \\ &= 22.25 + 22.5\varepsilon_1 + 12.5\varepsilon_2 + 25\varepsilon_3 + 6.25\varepsilon_4 - 12.5\varepsilon_5\end{aligned}$$

where $\varepsilon_i \in [-1,1]$ for $i = 3,4,5$ are newly generated noise symbols during the affine multiplications to evaluate the required result.

Here, the central value and the total deviation of the affine function $\hat{f}$ are $f_0 = 22.25$ and $d_f = 78.75$, respectively.

Converting the resulting affine solution into its interval bounds (as given in Section 3.3), we may have

$$\hat{f}(\hat{x},\hat{y}) = [22.25 - 78.75, 22.25 + 78.75] = [-56.5, 101].$$

Therefore, it may be observed from the above results that

$$f(x,y) \in \hat{f}(\hat{x},\hat{y}) \subset [f]([x],[y]),$$

where $\hat{f}(\hat{x},\hat{y})$ and $[f]([x],[y])$, respectively, are the functional values obtained by using the affine and interval arithmetic.

Hence (from Example 3.21), affine arithmetic may result in better enclosures as compared to the standard interval arithmetic.

3.7 EXERCISES

3.1. Find the central value and total deviation of the following affine representations:

(a) $\hat{x} = 20 + 5\varepsilon_1 - 2\varepsilon_2 + 7\varepsilon_3$,

(b) $\hat{y} = 6 - \varepsilon_1 + 10\varepsilon_2 - 4\varepsilon_4 - 12\varepsilon_5$,

(c) $\hat{z} = 150 + 19\varepsilon_3 - 28\varepsilon_5 + 38\varepsilon_6 + 15\varepsilon_7$.

3.2. Write the affine form representation of the affine multiplications $\hat{x}\hat{y}$, $\hat{y}\hat{z}$ and $\hat{z}\hat{x}$, where $\hat{x}$, $\hat{y}$, and $\hat{z}$ are the same as given in Exercise 3.1. Further, how many new noise symbols are generated in each of the above multiplications?

3.3. Find the interval bounds of the affine form representations given below:

(a) $\hat{a} = 50 - 23\varepsilon_1 - 23\varepsilon_2$,

(b) $\hat{b} = \varepsilon_1 - 5\varepsilon_2 + 10\varepsilon_3 + 20\varepsilon_4 - 16\varepsilon_5$,

(c) $\hat{c} = c - p\varepsilon_p - q\varepsilon_q$, where $c, p, q \in \mathbb{R}$.

3.4. Find the affine form representation of the intervals $[x] = [3.65, 5.7]$, $[y] = [1.5, 9.5]$, and $[z] = [15, 18]$. Thus, compute the affine solution of $5[x] + 8[y] - [z]$ and further reconvert the affine solution into its interval form.

3.5. Compute the affine solution of the nonlinear function $f(x, y) = x^2 + 2y^2 - 2xy$, where $\forall x \in [1, 3]$ and $\forall y \in [5, 7]$. Also, for this nonlinear function, check the efficacy of affine arithmetic compared to standard interval arithmetic.

3.8 REFERENCES

[1] Adusumilli, B. S. and Kumar, B. K., 2018. Modified affine arithmetic based continuation power flow analysis for voltage stability assessment under uncertainty. *IET Generation, Transmission and Distribution*, 12(18):4225–4232. DOI: 10.1049/iet-gtd.2018.5479.

[2] Akhmerov, R. R., 2005. Interval-affine Gaussian algorithm for constrained systems. *Reliable Computing*, 11(5):323–341. DOI: 10.1007/s11155-005-0040-5.

[3] Comba, J. L. D. and Stol, J., 1993. Affine arithmetic and its applications to computer graphics. In *Proc. of VI SIBGRAPI (Brazilian Symposium on Computer Graphics and Image Processing)*, pages 9–18. 39

[4] De Figueiredo, L. H. and Stolfi, J., 2004. Affine arithmetic: Concepts and applications. *Numerical Algorithms*, 37(1–4):147–158. DOI: 10.1023/b:numa.0000049462.70970.b6. 43

[5] Miyajima, S. and Kashiwagi, M., 2004. A dividing method utilizing the best multiplication in affine arithmetic. *IEICE Electronics Express*, 1(7):176–181. DOI: 10.1587/elex.1.176.

[6] Romero-Quete, D. and Cañizares, C. A., 2018. An affine arithmetic-based energy management system for isolated microgrids. *IEEE Transactions on Smart Grid*, 10(3):2989–2998. DOI: 10.1109/tsg.2018.2816403.

[7] Rump, S. M. and Kashiwagi, M., 2015. Implementation and improvements of affine arithmetic. *Nonlinear Theory and its Applications, IEICE*, 6(3):341–359. DOI: 10.1587/nolta.6.341.

[8] Skalna, I., 2009. Direct method for solving parametric interval linear systems with non-affine dependencies. In *International Conference on Parallel Processing and Applied Mathematics*, pages 485–494, Springer, Berlin, Heidelberg. DOI: 10.1007/978-3-642-14403-5_51. 43

[9] Skalna, I. and Hladík, M., 2017. A new algorithm for Chebyshev minimum-error multiplication of reduced affine forms. *Numerical Algorithms*, 76(4):1131–1152. DOI: 10.1007/s11075-017-0300-6. 46

[10] Stolfi, J. and De Figueiredo, L. H., 2003. An introduction to affine arithmetic. *Trends in Applied and Computational Mathematics*, 4(3):297–312. DOI: 10.5540/tema.2003.04.03.0297.

CHAPTER 4

Fuzzy-Affine Arithmetic

Fuzzy numbers and their arithmetic are a very powerful tool to handle uncertain parameters. By adopting the a-cut technique, fuzzy numbers can be parameterized and transformed into a family of intervals. All problems where the operands are in the form of different fuzzy numbers may be solved by using parametric fuzzy arithmetic. The parametric fuzzy arithmetic is based upon the concepts and properties of classic interval arithmetic. But the dependency problem or overestimation problem in standard interval arithmetic is a major hurdle that often leads to overestimation of the solution bounds. As such, fuzzy-affine arithmetic may be used to handle the fuzzy parameters more efficiently.

In this chapter, a-cut technique, basic terminologies and several operations of parametric fuzzy arithmetic have been incorporated. Further, the fuzzy-affine forms of different fuzzy numbers and fuzzy-affine arithmetic have been described. Finally, the efficiency and reliability of fuzzy-affine arithmetic have been shown by solving a few examples.

4.1 a-CUT OF FUZZY NUMBER

The "a-cut" of a fuzzy number $\tilde{F}$ is a crisp set given as

$$\tilde{F}(a) = \{x \in \mathbb{R} | \tilde{F}(x) \geq a\}. \tag{4.1}$$

Each a-cut $\tilde{F}(a)$ of the fuzzy number $\tilde{F}$ is a standard closed interval that depends upon the value of $a \in [0, 1]$ and may be represented as $\tilde{F}(a) = \left[\underline{F(a)}, \overline{F(a)}\right]$. Here, $\underline{F(a)}$ is the lower bound and $\overline{F(a)}$ is the upper bound of $\tilde{F}(a)$.

Note 4.1

The fuzzy set $\tilde{F}$ can be expressed uniquely and completely through the family of its a-cuts.

4.2 PARAMETRIC FORM OF FUZZY NUMBER

By adopting the a-cut technique, a fuzzy number may be parameterized and converted into an interval parametric form. If $\tilde{F}$ is a fuzzy number, then the parametric form of the fuzzy number may be denoted as $\tilde{F}(a)$ and is represented by

$$\tilde{F}(a) = \left[\underline{F(a)}, \overline{F(a)}\right], \quad \text{for} \quad a \in [0, 1]. \tag{4.2}$$

The lower bound $\left(\underline{F(a)}\right)$ and the upper bound $\left(\overline{F(a)}\right)$ of the interval parametric form of the fuzzy number $(\tilde{F}(a))$ must obey the below properties:

1. $\underline{F(a)}$ is the lower bound of the fuzzy number $\tilde{F}$, which is a left-bounded non-decreasing continuous function over $[0, 1]$.

2. $\overline{F(a)}$ is the upper bound of the fuzzy number $\tilde{F}$, which is a right-bounded non-increasing continuous function over $[0, 1]$.

3. For $a \in [0, 1]$, $\underline{F(a)} \leq \overline{F(a)}$.

The parametric form representations of different types of fuzzy numbers (viz. triangular fuzzy number (TFN) and trapezoidal fuzzy number (TrFN)) are illustrated in the following Sections 4.2.1 and 4.2.2.

4.2.1 PARAMETRIC FORM OF TFN

Let us consider a TFN $\tilde{T} = (t_1, t_2, t_3)$. By using the a-cut technique (for $a \in [0, 1]$), then TFN may be parameterized into a fuzzy interval form as follows:

$$\frac{x - t_1}{t_2 - t_1} = a \quad \text{and} \quad \frac{t_3 - x}{t_3 - t_2} = a. \tag{4.3}$$

That is,

$$\Rightarrow x = t_1 + a\,(t_2 - t_1) \quad \text{and} \quad \Rightarrow x = t_3 - a\,(t_3 - t_2). \tag{4.4}$$

Therefore, the interval parametric form of the TFN $\tilde{T} = (t_1, t_2, t_3)$ is obtained as

$$\tilde{T}(a) = \left[\underline{T(a)}, \overline{T(a)}\right] = [t_1 + a\,(t_2 - t_1)\,, t_3 - a\,(t_3 - t_2)]\,, \tag{4.5}$$

for $a \in [0, 1]$.

Note 4.2

1. Letting $a = 0$, the parametric interval form of the TFN (4.5) results in an interval quantity $[t_1, t_3]$, which is the outer enclosure of the TFN.

2. Substituting $a = 1$, the TFN (4.5) gives a crisp value t_2, which is the center of the TFN.

Example 4.3 Convert the TFN $\tilde{T} = (10, 13, 19)$ into its parametric interval form by adopting the a-cut technique.

Solution: Here, $t_1 = 10$, $t_2 = 13$, and $t_3 = 19$. Thus, by adopting the a-cut technique, the parametric interval form of the given TFN is computed as

$$\underline{T(a)} = t_1 + a\,(t_2 - t_1) = 10 + (13 - 10)a = 10 + 3a,$$
$$\overline{T(a)} = t_3 - a\,(t_3 - t_2) = 19 - (19 - 13)a = 19 - 6a.$$

Therefore, $\tilde{T}(a) = \left[\underline{T(a)}, \overline{T(a)}\right] = [10 + 3a, 19 - 6a]$, for $a \in [0, 1]$.

4.2.2 PARAMETRIC FORM OF TrFN

By utilizing the a-cut technique, a TrFN $\tilde{T}r = (r_1, r_2, r_3, r_4)$ may be parameterized and converted into the fuzzy interval form as follows:

$$\frac{x - r_1}{r_2 - r_1} = a \quad \text{and} \quad \frac{r_4 - x}{r_4 - r_3} = a. \tag{4.6}$$

Then, we have

$$\Rightarrow x = r_1 + a\,(r_2 - r_1) \quad \text{and} \quad \Rightarrow x = r_4 - a\,(r_4 - r_3)\,. \tag{4.7}$$

Therefore, the TrFN $\tilde{T}r = (r_1, r_2, r_3, r_4)$ is converted into its interval parametric form as

$$\tilde{T}r(a) = \left[\underline{Tr(a)}, \overline{Tr(a)}\right] = [r_1 + a\,(r_2 - r_1)\,, r_4 - a\,(r_4 - r_3)]\,, \tag{4.8}$$

for $a \in [0, 1]$.

Note 4.4

1. For $a = 0$, the parametric interval form of the TrFN (4.8) gives the outer interval bounds $[r_1, r_4]$ of the TrFN.
2. For $a = 1$, the parametric interval form of the TrFN (4.8) results in the interval $[r_2, r_3]$ of the TrFN.

Example 4.5 By using a-cut technique, evaluate the parametric interval form of a TrFN $\tilde{T}r = (5, 18, 23, 31)$.

Solution: Here, $r_1 = 5$, $r_2 = 18$, $r_3 = 23$, and $r_4 = 31$. Then, the parametric interval form of the given TrFN may be evaluated as follows:

$$\underline{Tr(a)} = r_1 + a\,(r_2 - r_1) = 5 + (18 - 5)a = 5 + 13a,$$
$$\overline{Tr(a)} = r_4 - a\,(r_4 - r_3) = 31 - (31 - 23)a = 31 - 8a.$$

Therefore, $\tilde{T}r(a) = \left[\underline{Tr(a)}, \overline{Tr(a)}\right] = [5 + 13a, 31 - 8a]$, for $a \in [0, 1]$.

4.3 BASIC TERMINOLOGIES OF PARAMETRIC FUZZY NUMBER

In this section, some of the basic terminologies associated with the interval parametric form of the fuzzy number are included and explained.

4.3.1 PARAMETRIC FUZZY CENTER ($\tilde{F}_c$)

The mid-point of the interval parametric form of the fuzzy number $\tilde{F}(a) = \left[\underline{F(a)}, \overline{F(a)}\right]$ is called its parametric fuzzy center and may be defined as

$$\tilde{F}_c = \frac{1}{2}\left(\underline{F(a)} + \overline{F(a)}\right). \tag{4.9}$$

4.3.2 PARAMETRIC FUZZY WIDTH ($\tilde{F}_w$)

The length between the lower bound and the upper bound of the interval parametric form of the fuzzy number $\tilde{F}(a) = \left[\underline{F(a)}, \overline{F(a)}\right]$ is referred to as its parametric fuzzy width and may be defined as

$$\tilde{F}_w = \overline{F(a)} - \underline{F(a)}. \tag{4.10}$$

4.3.3 PARAMETRIC FUZZY RADIUS ($\tilde{F}_\Delta$)

The half-length of the width of the interval parametric form of the fuzzy number $\tilde{F}(a) = \left[\underline{F(a)}, \overline{F(a)}\right]$ is called its parametric fuzzy radius and may be defined as

$$\tilde{F}_\Delta = \frac{1}{2}\left(\overline{F(a)} - \underline{F(a)}\right). \tag{4.11}$$

4.4 PARAMETRIC FUZZY ARITHMETIC

Let us consider the interval parametric form representation of two fuzzy numbers $\tilde{F}_1$ and $\tilde{F}_2$ as follows:

$$\tilde{F}_1 = \left[\underline{F_1(a)}, \overline{F_1(a)}\right] \quad \text{and} \quad \tilde{F}_2 = \left[\underline{F_2(a)}, \overline{F_2(a)}\right], \quad \text{for} \quad a \in [0, 1]. \tag{4.12}$$

Then, all the standard parametric fuzzy arithmetic (based on the concept of classic interval arithmetic) may be defined as

$$\tilde{F}_1 * \tilde{F}_1 = \{F_1 * F_2 | F_1 \in F_1(a), F_2 \in F_2(a) \quad \text{and} \quad a \in [0, 1]\}, \tag{4.13}$$

where "$*$" stands for all the binary operations viz. $\{+, -, \cdot, /\}$. Keeping this in view, the binary operations of the parametric fuzzy arithmetic are discussed below.

4.4.1 PARAMETRIC FUZZY ADDITION

$$\tilde{F}_1 + \tilde{F}_2 = \left[\underline{F_1(a)} + \underline{F_2(a)}, \overline{F_1(a)} + \overline{F_2(a)}\right]. \tag{4.14}$$

4.4.2 PARAMETRIC FUZZY SUBTRACTION

$$\tilde{F}_1 - \tilde{F}_2 = \left[\underline{F_1(a)} - \overline{F_2(a)}, \overline{F_1(a)} - \underline{F_2(a)}\right]. \tag{4.15}$$

4.4.3 PARAMETRIC FUZZY SCALAR MULTIPLICATION

$$k \cdot \tilde{F}_1 = \begin{cases} \left[k \cdot \underline{F_1(a)}, k \cdot \overline{F_1(a)}\right], & k \geq 0 \\ \left[k \cdot \overline{F_1(a)}, k \cdot \underline{F_1(a)}\right] & k < 0, \end{cases} \quad \text{for} \quad k \in \mathbb{R}. \tag{4.16}$$

4.4.4 PARAMETRIC FUZZY MULTIPLICATION

$$\tilde{F}_1 \cdot \tilde{F}_2 = \left[\min\left\{\Gamma\left(\tilde{F}_1, \tilde{F}_2\right)\right\}, \ \max\left\{\Gamma\left(\tilde{F}_1, \tilde{F}_2\right)\right\}\right], \tag{4.17a}$$

where

$$\Gamma\left(\tilde{F}_1, \tilde{F}_2\right) = \left\{\underline{F_1(a)} \cdot \underline{F_2(a)}, \underline{F_1(a)} \cdot \overline{F_2(a)}, \overline{F_1(a)} \cdot \underline{F_2(a)}, \overline{F_1(a)} \cdot \overline{F_2(a)}\right\}. \tag{4.17b}$$

4.4.5 PARAMETRIC FUZZY RECIPROCAL

$$\frac{1}{\tilde{F}_1} = \begin{cases} \left[\dfrac{1}{\overline{F_1(a)}}, \dfrac{1}{\underline{F_1(a)}}\right], & 0 \notin \tilde{F}_1 \\ (-\infty, \infty), & 0 \in \tilde{F}_1. \end{cases} \tag{4.18}$$

4.4.6 PARAMETRIC FUZZY DIVISION

$$\frac{\tilde{F}_1}{\tilde{F}_2} = \tilde{F}_1 \cdot \frac{1}{\tilde{F}_2} = \begin{cases} \left[\underline{F_1(a)}, \overline{F_1(a)}\right] \cdot \left[\dfrac{1}{\overline{F_2(a)}}, \dfrac{1}{\underline{F_2(a)}}\right] & 0 \notin \tilde{F}_2 \\ (-\infty, \infty) & 0 \in \tilde{F}_2. \end{cases} \tag{4.19}$$

4.4.7 PARAMETRIC FUZZY POWER

- If $m > 0$ is an odd integer, then

$$\left(\tilde{F}_1\right)^m = \left[\underline{F_1(a)^m}, \overline{F_1(a)^m}\right]. \tag{4.20a}$$

- If $m > 0$ is an even integer, then

$$\left(\tilde{F}_1\right)^m = \begin{cases} \left[\underline{F_1(a)^m}, \overline{F_1(a)^m}\right], & \tilde{F}_1 > 0 \\ \left[\overline{F_1(a)^m}, \underline{F_1(a)^m}\right], & \tilde{F}_1 < 0 \\ \left[0, \max\left\{\underline{F_1(a)^m}, \overline{F_1(a)^m}\right\}\right], & 0 \in \tilde{F}_1. \end{cases} \tag{4.20b}$$

Example 4.6 Add and multiply two TFNs $\tilde{T}_1 = (2, 8, 17)$ and $\tilde{T}_2 = (-22, -14, -4)$ by adopting the parametric fuzzy arithmetic and draw the fuzzy solution plot.

Solution: First, we have to convert the given TFNs into its interval parametric form representation by using a-cut technique. Thus, we have

$$\begin{aligned} \tilde{T}_1(a) &= [2 + (8-2)a, 17 - (17-8)a] = [2 + 6a, 17 - 9a] \qquad \text{and} \\ \tilde{T}_2(a) &= [-22 + (-14+22)a, -4 - (-4+14)a] \\ &= [-22 + 8a, -4 - 10a], \quad \text{for} \quad a \in [0, 1]. \end{aligned}$$

The sum of two given TFNs (from Eq. (4.14)) is

$$\begin{aligned} \tilde{T}_1(a) + \tilde{T}_2(a) &= \left[\underline{T_1(a)} + \underline{T_2(a)}, \overline{T_1(a)} + \overline{T_2(a)}\right] \\ &= [(2 + 6a) + (-22 + 8a), (17 - 9a) + (-4 - 10a)] \\ &= [-20 + 14a,\ 13 - 19a]. \end{aligned}$$

Again, the product of the TFNs (from Section 4.4.4) is computed as follows:

$$\begin{aligned} \Gamma\left(\tilde{T}_1, \tilde{T}_2\right) &= \left\{\underline{T_1(a)} \cdot \underline{T_2(a)}, \underline{T_1(a)} \cdot \overline{T_2(a)}, \overline{T_1(a)} \cdot \underline{T_2(a)}, \overline{T_1(a)} \cdot \overline{T_2(a)}\right\}, \\ &= \{(2 + 6a)(-22 + 8a), (2 + 6a)(-4 - 10a), \\ &\qquad (17 - 9a)(-22 + 8a), (17 - 9a)(-4 - 10a)\} \\ &= \{(-44 - 116a + 48a^2), (-8 - 44a - 60a^2), (-374 + 334a - 72a^2), \\ &\qquad (-68 - 134a + 90a^2)\} \\ \tilde{T}_1 \cdot \tilde{T}_2 &= \left[\min\left\{\Gamma(\tilde{T}_1, \tilde{T}_2)\right\}, \max\left\{\Gamma\left(\tilde{T}_1, \tilde{T}_2\right)\right\}\right] \\ &= \left[(-374 + 334a - 72a^2), (-8 - 44a - 60a^2)\right]. \end{aligned}$$

Therefore, the resulting TFN solutions are found to be $\tilde{T}_1 + \tilde{T}_2 = [-20 + 14a, 13 - 19a]$ and $\tilde{T}_1 \cdot \tilde{T}_2 = \left[(-374 + 334a - 72a^2), (-8 - 44a - 60a^2)\right]$, for $a \in [0, 1]$. Its solution plots are depicted in Figs. 4.1–4.2.

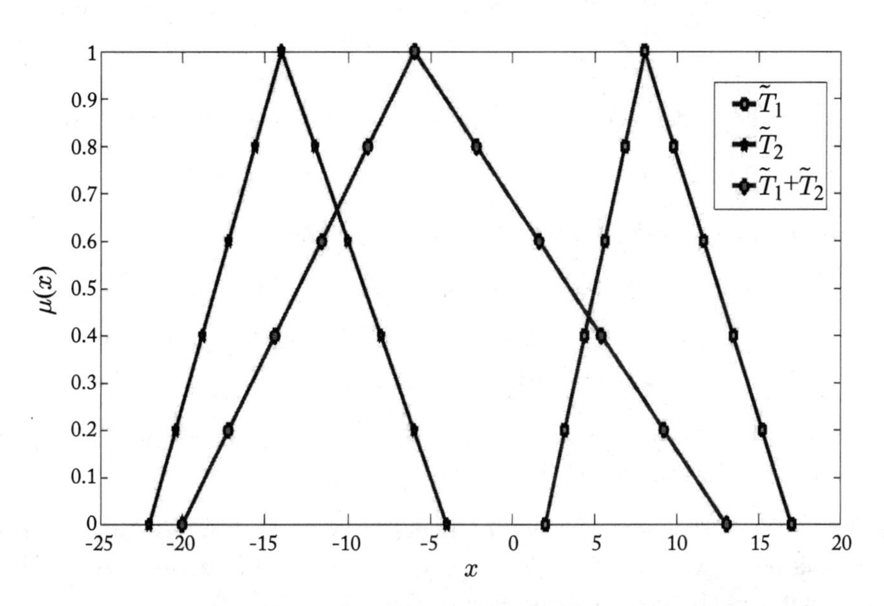

Figure 4.1: Sum of the two TFNs of Example 4.6 by using parametric fuzzy arithmetic.

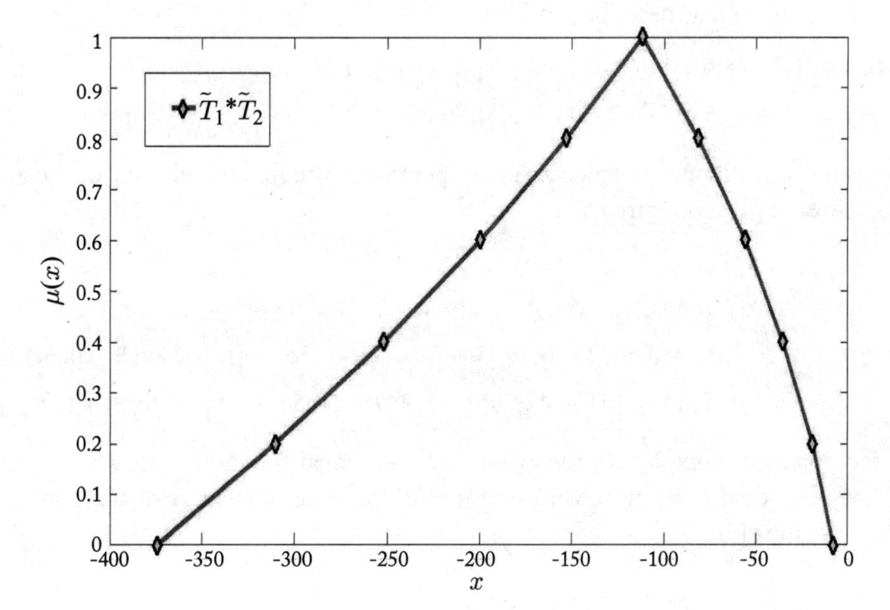

Figure 4.2: Product of the two TFNs of Example 4.6 by using parametric fuzzy arithmetic.

Example 4.7 Subtract a TrFN $\tilde{T}r = (-5, 0, 5, 10)$ from itself by using parametric fuzzy arithmetic.

Solution: By using a-cut technique, the given TrFN $\tilde{T}r = (-5, 0, 5, 10)$ is parameterized as

$$\tilde{T}r(a) = [-5 + 5a, 10 - 5a], \quad \text{for} \quad a \in [0, 1].$$

Then, subtracting $\tilde{T}r = (-5, 0, 5, 10)$ from itself, we may have

$$\begin{aligned}\tilde{T}r - \tilde{T}r &= [-5 + 5a, 10 - 5a] - [-5 + 5a, 10 - 5a] \\ &= [(-5 + 5a) - (10 - 5a), (10 - 5a) - (-5 + 5a)] \\ &= [-15 + 10a, 15 - 10a].\end{aligned}$$

It may be noted that, by substituting $a = 0$, the outer enclosure of the resulting interval is $[-15, 15]$, whose width is twice the width of the operand.

Example 4.8 Using parametric fuzzy arithmetic, compute the fuzzy solution plot of a fuzzy nonlinear function $\tilde{f}(\tilde{x}, \tilde{y}) = 2\tilde{x}^2 + 5\tilde{y}^2 - 3\tilde{x}\tilde{y} - \tilde{x}$, where the variables are in the form of TrFNs such that $\tilde{x} = (1, 1.5, 2.5, 3)$ and $\tilde{y} = (5, 5.5, 6.5, 7)$.

Solution: First, we have to parameterize (using a-cut) the given TrFNs $\tilde{x} = (1, 1.5, 2.5, 3)$ and $\tilde{y} = (5, 5.5, 6.5, 7)$, as described in Section 4.2.2, as follows:

$$\begin{aligned}\tilde{x}(a) &= [1 + (1.5 - 1)a, 3 - (3 - 2.5)a] = [1 + 0.5a, 3 - 0.5a] \quad \text{and} \\ \tilde{y} &= [5 + (5.5 - 5)a, 7 - (7 - 6.5)a] = [5 + 0.5a, 7 - 0.5a], \quad \text{for} \quad a \in [0, 1].\end{aligned}$$

Thus, by using the different parametric fuzzy operations, the functional value of the given nonlinear function may be computed as

$$\begin{aligned}\tilde{f}(\tilde{x}, \tilde{y}) &= 2\tilde{x}^2 + 5\tilde{y}^2 - 3\tilde{x}\tilde{y} - \tilde{x} \\ &= 2[1 + 0.5a, 3 - 0.5a]^2 + 5[5 + 0.5a, 7 - 0.5a]^2 \\ &\quad - 3[1 + 0.5a, 3 - 0.5a][5 + 0.5a, 7 - 0.5a] - [1 + 0.5a, 3 - 0.5a] \\ \Rightarrow \tilde{f}(\tilde{x}, \tilde{y}) &= \left[61 + 42.5a + a^2, 247 - 50.5a + a^2\right], \quad \text{for} \quad a \in [0, 1].\end{aligned}$$

Finally, the fuzzy solution plot for the given fuzzy nonlinear function is depicted in Fig. 4.3.

It may be noted that, by substituting $a = 0$, the outer enclosure of the nonlinear fuzzy function $\tilde{f}(\tilde{x}, \tilde{y})$ is $[61, 247]$.

4.5 FUZZY-AFFINE FORM

In standard fuzzy arithmetic (or parametric fuzzy arithmetic), all the operands are assumed to be independent of each other. But when they are partially or completely dependent on each other,

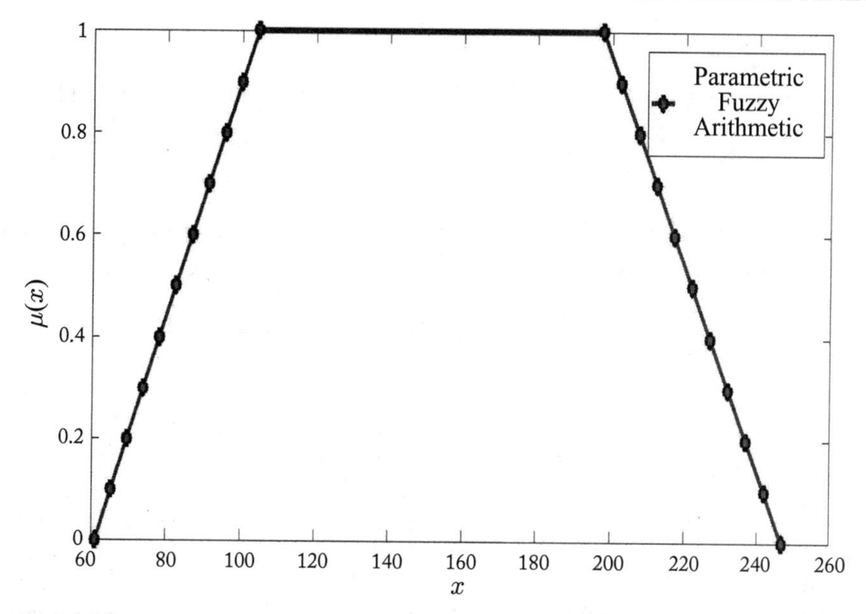

Figure 4.3: TrFN solution of Example 4.8 using parametric fuzzy arithmetic.

the standard fuzzy arithmetic results in a wider range. This is known as the "dependency problem" or "overestimation problem." In this regard, fuzzy-affine arithmetic is developed to overcome the overestimation problem in the case of standard fuzzy arithmetic. To perform fuzzy-affine arithmetic, first of all, we have to transform the fuzzy numbers into their respective fuzzy-affine forms having different parameters viz. fuzzy parameter (a) and noise symbols (ε_i). Keeping this in view, fuzzy-affine forms of different fuzzy numbers are in the below sections.

4.5.1 FUZZY-AFFINE FORM OF TFN

Let us consider a TFN $\tilde{T} = (t_1, t_2, t_3)$. From Section 4.2.1, by using the a-cut technique, the given TFN is converted into an interval parametric form as follows:

$$\tilde{T}(a) = \left[\underline{T(a)}, \overline{T(a)}\right] = [t_1 + a(t_2 - t_1), t_3 - a(t_3 - t_2)]. \tag{4.21}$$

Then, the central value and half-width (or radius) of the above interval parametric form may be computed as

$$\begin{aligned}\tilde{T}_c(a) &= \frac{1}{2}\left(\underline{T(a)} + \overline{T(a)}\right) = \frac{1}{2}\{(t_1 + a(t_2 - t_1)) + (t_3 - a(t_3 - t_2))\} \\ &= \frac{1}{2}\{(t_1 + t_3) + a(2t_2 - t_1 - t_3)\},\end{aligned} \tag{4.22a}$$

and

$$\begin{aligned}\tilde{T}_{\Delta}(a) &= \frac{1}{2}\left(\overline{T(a)} - \underline{T(a)}\right) = \frac{1}{2}\{(t_3 - a(t_3 - t_2)) - (t_1 + a(t_2 - t_1))\} \\ &= \frac{1}{2}\{(t_3 - t_1) + a(t_1 - t_3)\}.\end{aligned} \tag{4.22b}$$

Therefore, according to Chapter 3, the above parametric form further may be transformed into an affine form representation as

$$\hat{T}(a, \varepsilon_t) = \tilde{T}_c(a) + \tilde{T}_{\Delta}(a)\varepsilon_t, \quad \text{for} \quad \varepsilon_t \in [-1, 1]. \tag{4.23}$$

That is,

$$\hat{T}(a, \varepsilon_t) = \frac{1}{2}\{(t_1 + t_3) + a(2t_2 - t_1 - t_3)\} + \frac{1}{2}\{(t_3 - t_1) + a(t_1 - t_3)\}\varepsilon_t, \tag{4.24}$$

for $a \in [0, 1]$ and $\varepsilon_t \in [-1, 1]$. Here, ε_t is the noise symbol for the fuzzy-affine form representation (4.24).

Example 4.9 Find the fuzzy-affine form of a TFN $\tilde{T} = (-39, -25, -8)$ by utilizing the formula (4.24).

Solution: Here, $t_1 = -39$, $t_2 = -25$, and $t_3 = -8$. Then, according to the formula (4.24), the fuzzy-affine form of the given TFN is

$$\begin{aligned}\hat{T}(a, \varepsilon_1) &= \frac{1}{2}\{(t_1 + t_3) + a(2t_2 - t_1 - t_3)\} + \frac{1}{2}\{(t_3 - t_1) + a(t_1 - t_3)\}\varepsilon_1 \\ &= \frac{1}{2}\{(-39 - 8) + a(2(-25) + 39 + 8)\} + \frac{1}{2}\{(-8 + 39) + a(-39 + 8)\}\varepsilon_1 \\ &= \frac{1}{2}\{-47 + a(-3)\} + \frac{1}{2}\{31 + a(-31)\}\varepsilon_1 \\ &= (-23.5 - 1.5a) + (15.5 - 15.5a)\varepsilon_1,\end{aligned}$$

where ε_1 is the noise symbol of the fuzzy-affine form representation that lies in the interval $[-1, 1]$.

Hence, the fuzzy-affine form of the given TFN is $\hat{T}(a, \varepsilon_1) = (-23.5 - 1.5a) + (15.5 - 15.5a)\varepsilon_1$, for $a \in [0, 1]$ and $\varepsilon_1 \in [-1, 1]$.

4.5.2 FUZZY-AFFINE FORM OF TrFN

As given in Section 4.2.2, the interval parametric form of a TrFN $\tilde{T}r = (r_1, r_2, r_3, r_4)$ by using the a- cut technique is written as:

$$\tilde{T}r(a) = \left[\underline{Tr(a)}, \overline{Tr(a)}\right] = [r_1 + a(r_2 - r_1), r_4 - a(r_4 - r_3)]. \tag{4.25}$$

Similarly, the central value and radius of the interval parametric form (4.25) is

$$\tilde{Tr}_c(a) = \frac{1}{2}\left(\underline{Tr(a)} + \overline{Tr(a)}\right) = \frac{1}{2}\{(r_1 + a\,(r_2 - r_1)) + (r_4 - a\,(r_4 - r_3))\}$$
$$= \frac{1}{2}\{(r_1 + r_4) + a\,((r_2 + r_3) - (r_1 + r_4))\}, \tag{4.26a}$$

and

$$\tilde{Tr}_\Delta(a) = \frac{1}{2}\left(\overline{Tr(a)} - \underline{Tr(a)}\right) = \frac{1}{2}\{(r_4 - a\,(r_4 - r_3)) - (r_1 + a\,(r_2 - r_1))\}$$
$$= \frac{1}{2}\{(r_4 - r_1) + a\,((r_1 + r_3) - (r_2 + r_4))\}. \tag{4.26b}$$

Hence, the parametric form (4.25) is converted into its affine form representation (as given in Chapter 3) as follows:

$$\hat{Tr}\,(a, \varepsilon_r) = \tilde{Tr}_c(a) + \tilde{Tr}_\Delta(a)\varepsilon_r, \quad \text{for} \quad \varepsilon_r \in [-1, 1]. \tag{4.27}$$

That is,

$$\hat{Tr}\,(a, \varepsilon_r) = \frac{1}{2}\{(r_1 + r_4) + a\,((r_2 + r_3) - (r_1 + r_4))\}$$
$$+ \frac{1}{2}\{(r_4 - r_1) + a\,((r_1 + r_3) - (r_2 + r_4))\}\,\varepsilon_r, \tag{4.28}$$

for $a \in [0, 1]$ and $\varepsilon_r \in [-1, 1]$, where ε_r is the noise symbol for the fuzzy-affine form representation (4.28).

Example 4.10 Let us consider a TrFN $\tilde{Tr} = (-65, -40, -25, 10)$. Transform the given TrFN into its fuzzy-affine form representation.

Solution: Here, $r_1 = -65$, $r_2 = -40$, $r_3 = -25$, and $r_4 = 10$. First, we are going to convert the given TrFN into its interval parametric form as follows:

$$\tilde{Tr}(a) = \left[\underline{Tr(a)}, \overline{Tr(a)}\right] = [r_1 + a\,(r_2 - r_1)\,, r_4 - a\,(r_4 - r_3)]$$
$$= [-65 + a(-40 + 65), 10 - a(10 + 25)] = [-65 + 25a, 10 - 35a]\,, \quad \text{for} \quad a \in [0, 1].$$

The central value and radius of the above parametric form is

$$\tilde{Tr}_c(a) = \frac{1}{2}\left(\underline{Tr(a)} + \overline{Tr(a)}\right) = \frac{1}{2}\{(-65 + 25a) + (10 - 35a)\} = -27.5 - 5a \quad \text{and}$$
$$\tilde{Tr}_\Delta(a) = \frac{1}{2}\left(\overline{Tr(a)} - \underline{Tr(a)}\right) = \frac{1}{2}\{(10 - 35a) - (-65 + 25a)\} = 37.5 - 30a.$$

Then, the interval parametric form further may be transformed into the fuzzy-affine form as

$$\hat{T}r\,(a,\varepsilon_2) = \tilde{T}r_c(a) + \tilde{T}r_\Delta(a)\varepsilon_2$$
$$= (-27.5 - 5a) + (37.5 - 30a)\varepsilon_2, \quad \text{for} \quad a \in [0,1] \quad \text{and} \quad \varepsilon_2 \in [-1,1].$$

Here, $\varepsilon_2 \in [-1,1]$ is the noise symbol of the fuzzy-affine form representation.

4.6 FUZZY-AFFINE ARITHMETIC OPERATIONS

Let us consider two fuzzy numbers $\tilde{F}_1$ and $\tilde{F}_2$. By adopting the procedure given in Section 4.5, these two fuzzy numbers may be transformed into respective fuzzy-affine forms having different noise symbols (ε_1 and ε_2) such as

$$\tilde{F}_1 \approx \hat{F}_1\,(a,\varepsilon_1) \quad \text{and} \quad \tilde{F}_2 \approx \hat{F}_2\,(a,\varepsilon_2), \tag{4.29}$$

for $a \in [0,1]$ and $\varepsilon_1, \varepsilon_2 \in [-1,1]$.

Here, the above fuzzy-affine forms have several parameters viz. fuzzy parameter (a) and different noise symbols (ε_i for $i = 1,2$). In this regard, all the operations regarding fuzzy-affine arithmetic can be defined as follows:

$$\hat{R}\,(a,\varepsilon_1,\varepsilon_2) = \hat{F}_1\,(a,\varepsilon_1) * \hat{F}_2\,(a,\varepsilon_2), \tag{4.30}$$

where $\hat{R}(a,\varepsilon_1,\varepsilon_2)$ is the fuzzy-affine form of the resulting solution having all the existing noise symbols and some newly generated noise symbols.

Hence, the interval parametric solution may be obtained as

$$\tilde{R} = \tilde{F}_1 * \tilde{F}_2 = \left[\underline{R(a)}, \overline{R(a)}\right], \quad \text{and} \quad \forall a \in [0,1], \tag{4.31}$$

where

$$\underline{R(a)} = \min_{\varepsilon_1,\varepsilon_2 \in [-1,1]} \hat{R}\,(a,\varepsilon_1,\varepsilon_2) \quad \text{and} \quad \overline{R(a)} = \max_{\varepsilon_1,\varepsilon_2 \in [-1,1]} \hat{R}\,(a,\varepsilon_1,\varepsilon_2). \tag{4.32}$$

Therefore, the final resulting fuzzy solution may be computed by varying the fuzzy parameter (a) from 0 to 1.

The detailed procedures to apply all the operations of fuzzy-affine arithmetic are described in the following examples (Example 4.11–4.15).

Example 4.11 By using fuzzy-affine arithmetic, find the sum and product of two TFNs $\tilde{T}_1 = (-50,-38,-9)$ and $\tilde{T}_2 = (11,44,62)$. Further, plot the fuzzy triangular solutions.

Solution: Parameterizing the given two TFNs by using a-cut technique, the fuzzy parametric forms are found as

$$\tilde{T}_1(a) = [-50 + 12a, -9 - 29a] \quad \text{and} \quad \tilde{T}_2(a) = [11 + 33a, 62 - 18a], \quad \text{for} \quad a \in [0,1].$$

Further, the above interval parametric form may be converted into a fuzzy-affine form by adopting the procedure given in Section 4.5.1 as follows:

$$\hat{T}_1\,(a,\varepsilon_1) = (-29.5 - 8.5a) + (20.5 - 20.5a)\varepsilon_1 \quad \text{and}$$
$$\tilde{T}_2\,(a,\varepsilon_2) = (36.5 + 7.5a) + (25.5 - 25.5a)\varepsilon_2,$$

where $a \in [0,1]$ and $\varepsilon_1, \varepsilon_2 \in [-1,1]$.

Suppose, $\hat{S}$ and $\hat{P}$ are the respective sum and product of the given TFNs. Then by fuzzy-affine arithmetic, we may have

$$\begin{aligned}
\hat{S} &= \hat{T}_1\,(a,\varepsilon_1) + \hat{T}_2\,(a,\varepsilon_2) \\
&= \{(-29.5 - 8.5a) + (20.5 - 20.5a)\varepsilon_1\} + \{(36.5 + 7.5a) + (25.5 - 25.5a)\varepsilon_2\}, \\
&\Rightarrow \hat{S} = (7 - a) + (20.5 - 20.5a)\varepsilon_1 + (25.5 - 25.5a)\varepsilon_2, \\
&\quad \text{for} \quad a \in [0,1] \quad \text{and} \quad \varepsilon_1, \varepsilon_2 \in [-1,1],
\end{aligned}$$

and

$$\begin{aligned}
\hat{P} &= \hat{T}_1\,(a,\varepsilon_1) \cdot \hat{T}_2\,(a,\varepsilon_2) \\
&= \{(-29.5 - 8.5a) + (20.5 - 20.5a)\varepsilon_1\} \cdot \{(36.5 + 7.5a) + (25.5 - 25.5a)\varepsilon_2\}, \\
&\Rightarrow \hat{P} = \left(-1076 - 531.5a - 63.75a^2\right) + \left(748.25 - 594.5a - 153.75a^2\right)\varepsilon_1 \\
&\quad + \left(-752.25 + 535.5a + 216.75a^2\right)\varepsilon_2 + \left(522.75 - 1045.5a + 522.75a^2\right)\varepsilon_3,
\end{aligned}$$

where $a \in [0,1]$ and $\varepsilon_i \in [-1,1]$ for $i = 1,2,3$. Here, $\varepsilon_3 = \varepsilon_1 \cdot \varepsilon_2$ is the new noise symbol generated during the fuzzy-affine multiplication.

Thus, all the resulting TFN solution plots are as depicted in Figs. 4.4–4.5.

Note 4.12

In fuzzy arithmetic operations, where all the operands are in the form of a fuzzy number, the resulting solution may or may not be a fuzzy number.

Example 4.13 Consider the TrFN as given in Example 4.7 (that is $\tilde{T}r = (-5, 0, 5, 10)$). Subtract this TrFN from itself by using fuzzy-affine arithmetic.

Solution: From Example 4.7, the parametric form of the given TrFN (by using a-cut) is

$$\tilde{T}r(a) = [-5 + 5a, 10 - 5a], \quad \text{for} \quad a \in [0,1].$$

Now, we have to convert it into a fuzzy-affine form with the noise symbol ε_1 as mentioned in Section 4.5. Thus,

$$\hat{T}r\,(a,\varepsilon_1) = 2.5 + (7.5 - 5a)\varepsilon_1, \quad \text{for} \quad a \in [0,1] \quad \text{and} \quad \varepsilon_1 \in [-1,1].$$

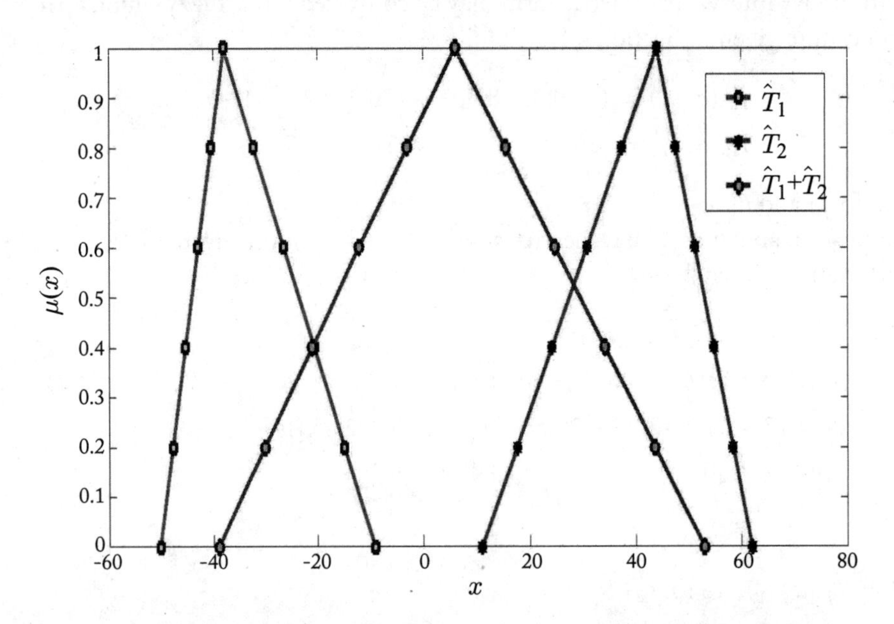

Figure 4.4: Sum of the two TFNs of Example 4.11 by using fuzzy-affine arithmetic.

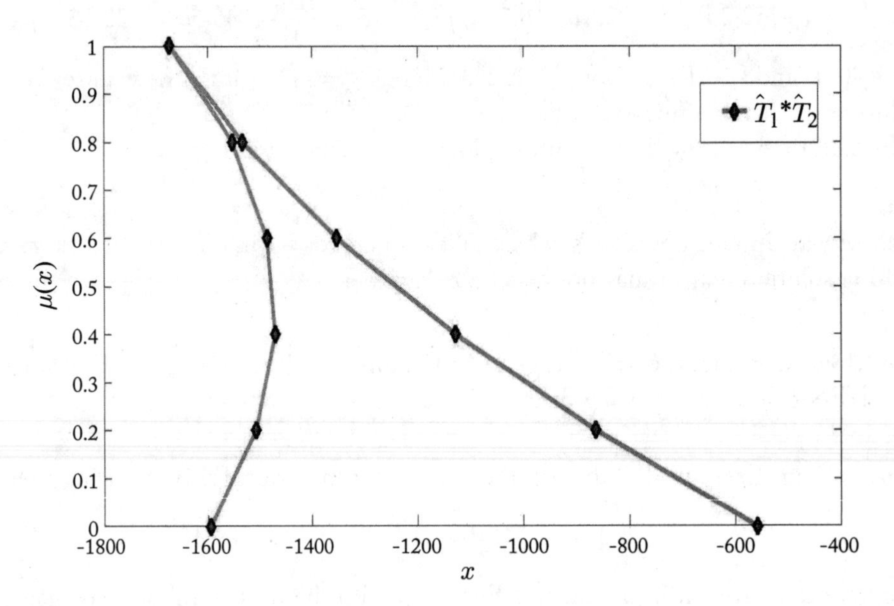

Figure 4.5: Product of the two TFNs of Example 4.11 by using fuzzy-affine arithmetic.

Hence, subtracting the TrFN from itself, we may have

$$\begin{aligned}\tilde{T}r - \tilde{T}r &= \hat{T}r\,(a, \varepsilon_1) - \hat{T}r\,(a, \varepsilon_1) \\ &= \{2.5 + (7.5 - 5a)\varepsilon_1\} - \{2.5 + (7.5 - 5a)\varepsilon_1\} = 0.\end{aligned}$$

Therefore, $\tilde{T}r - \tilde{T}r = 0$.

Note 4.14

It can be shown from Example 4.7 that subtraction of any fuzzy number from itself by using standard parametric fuzzy arithmetic results in a comparatively wider outer enclosure instead of being zero. This is one of the major reasons behind the overestimation problem. But in the case of fuzzy-affine arithmetic, the subtraction of a fuzzy number from itself results in zero (from Example 4.13). Thus, fuzzy-affine arithmetic may overcome the overestimation problem that occurred in the case of standard fuzzy arithmetic.

Example 4.15 Let us consider the same fuzzy nonlinear function as given in Example 4.8 (that is, $\tilde{f}\,(\tilde{x}, \tilde{y}) = 2\tilde{x}^2 + 5\tilde{y}^2 - 3\tilde{x}\tilde{y} - \tilde{x}$) with the same variables such as $\tilde{x} = (1, 1.5, 2.5, 3)$ and $\tilde{y} = (5, 5.5, 6.5, 7)$. Evaluate its functional value by using fuzzy-affine arithmetic and also plot its solution graph.

Solution: As found in Example 4.8, the interval parametric forms (by using a-cut technique) of the given variables are

$$\tilde{x}(a) = [1 + 0.5a, 3 - 0.5a] \quad \text{and} \quad \tilde{y} = [5 + 0.5a, 7 - 0.5a], \quad \text{for} \quad a \in [0, 1].$$

Further, the above interval parametric form may be converted into its fuzzy-affine forms (as given in Section 4.5) with respective noise symbols ε_1 and ε_2 as follows:

$$\hat{x}\,(a, \varepsilon_1) = 2 + (1 - 0.5a)\varepsilon_1 \quad \text{and} \quad \hat{y}\,(a, \varepsilon_2) = 6 + (1 - 0.5a)\varepsilon_2,$$

where $a \in [0, 1]$ and $\varepsilon_i = [-1, 1]$ for $i = 1, 2$.

Thus, by adopting fuzzy-affine arithmetic, the functional value of the given fuzzy nonlinear function is

$$\begin{aligned}\hat{f}\,(\hat{x}, \hat{y}) &= 2\hat{x}^2 + 5\hat{y}^2 - 3\hat{x}\hat{y} - \hat{x} \\ &= 2\,(2 + (1 - 0.5a)\varepsilon_1)^2 + 5\,(6 + (1 - 0.5a)\varepsilon_2)^2 \\ &\quad - 3\,(2 + (1 - 0.5a)\varepsilon_1)\,(6 + (1 - 0.5a)\varepsilon_2) - (2 + (1 - 0.5a)\varepsilon_1) \\ &\Rightarrow \hat{f}\,(\hat{x}, \hat{y}, \hat{z}) = 186 - 11(1 - 0.5a)\varepsilon_1 + 54(1 - 0.5a)\varepsilon_2 + 2(1 - 0.5a)^2\varepsilon_3 \\ &\quad + 5(1 - 0.5a)^2\varepsilon_4 - 3(1 - 0.5a)^2\varepsilon_5,\end{aligned}$$

where $a \in [0, 1]$ and $\varepsilon_i = [-1, 1]$ for $i = 1, \ldots, 5$. Here, $\varepsilon_3 = (\varepsilon_1)^2$, $\varepsilon_4 = (\varepsilon_2)^2$ and $\varepsilon_5 = \varepsilon_1\varepsilon_2$ are noise symbols newly generated during all the fuzzy-affine operations.

Finally, the required trapezoidal fuzzy solution plot is as depicted in Fig. 4.6.

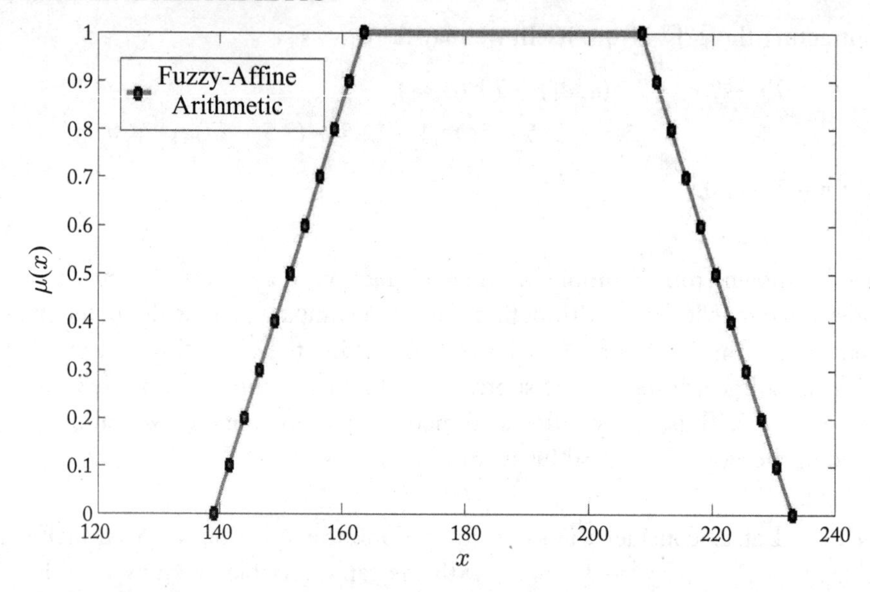

Figure 4.6: TrFN solution of Example 4.15 using fuzzy-affine arithmetic.

4.7 EFFICACY OF FUZZY-AFFINE ARITHMETIC

As mentioned above, fuzzy-affine arithmetic can overcome the overestimation problem which occurred during standard fuzzy arithmetic. In this section, a few examples (Examples 4.16–4.17) have been worked out to show the reliability and efficacy of the fuzzy-affine arithmetic.

Example 4.16 Let us consider a fuzzy linear function $\tilde{f}(\tilde{x}, \tilde{y}, \tilde{z}) = \alpha\tilde{x} + \beta\tilde{y} + \gamma\tilde{z} + \delta$, where all the consistent variables are taken in the form of TrFNs such that $\tilde{x} = (-5, -4, -2, -1)$, $\tilde{y} = (1, 1.25, 1.75, 2)$, and $\tilde{z} = (3, 4, 5, 6)$. Further, all the coefficients of the linear function are crisp numbers viz. $\alpha = 10$, $\beta = -6$, $\gamma = 2$, and $\delta = -7$. Perform both parametric fuzzy arithmetic and fuzzy-affine arithmetic to compute the functional value of the given fuzzy linear function and show the efficacy of fuzzy-affine arithmetic through a solution plot.

Solution: To perform parametric fuzzy arithmetic, let us convert the TrFNs into the parametric interval form (by using a-cut technique) given as follows:

$$\tilde{x}(a) = [-5 + a, -1 - a], \quad \tilde{y}(a) = [1 + 0.25a, 2 - 0.25a]$$
$$\text{and} \quad \tilde{z}(a) = [3 + a, 6 - a], \quad \text{for} \quad a \in [0, 1].$$

Then, the functional value of the given fuzzy linear function by using parametric fuzzy arithmetic is

$$\begin{aligned}\tilde{f}(\tilde{x}, \tilde{y}, \tilde{z}) &= \alpha\tilde{x} + \beta\tilde{y} + \gamma\tilde{z} + \delta \\ &= 10[-5 + a, -1 - a] + (-6)[1 + 0.25a, 2 - 0.25a] + 2[3 + a, 6 - a] + (-7) \\ &\Rightarrow \tilde{f}(\tilde{x}, \tilde{y}, \tilde{z}) = [-63 + 13.5a, -11 - 13.5a], \quad \text{for} \quad a \in [0, 1].\end{aligned}$$

Further, to perform fuzzy-affine arithmetic, we have to transfer the given variables into their respective fuzzy-affine forms as follows:

$$\hat{x}(a, \varepsilon_1) = -3 + (2 - a)\varepsilon_1, \hat{y}(a, \varepsilon_2) = 1.5 + (0.5 - 0.25a)\varepsilon_2$$
$$\text{and} \quad \hat{z}(a, \varepsilon_3) = 4.5 + (1.5 - a)\varepsilon_3,$$

where $a \in [0, 1]$ and $\varepsilon_i = [-1, 1]$ for $i = 1, 2, 3$ are the respective noise symbols of the above fuzzy-affine forms.

Thus, by adopting fuzzy-affine arithmetic, the functional value of the given fuzzy linear function is

$$\begin{aligned}\hat{f}(\hat{x}, \hat{y}, \hat{z}) &= \alpha\hat{x} + \beta\hat{y} + \gamma\hat{z} + \delta \\ &= 10(-3 + (2 - a)\varepsilon_1) + (-6)(1.5 + (0.5 - 0.25a)\varepsilon_2) \\ &\quad + 2(4.5 + (1.5 - a)\varepsilon_3) + (-7) \\ &\Rightarrow \hat{f}(\hat{x}, \hat{y}, \hat{z}) = -37 + 10(2 - a)\varepsilon_1 - 6(0.5 - 0.25a)\varepsilon_2 + 2(1.5 - a)\varepsilon_3,\end{aligned}$$

where $a \in [0, 1]$ and $\varepsilon_i = [-1, 1]$ for $i = 1, 2, 3$.

Now, the corresponding fuzzy (TrFN) functional values of the given nonlinear function with respect to parametric fuzzy arithmetic $\left(\tilde{f}(\tilde{x}, \tilde{y}, \tilde{z})\right)$ and fuzzy-affine arithmetic $\left(\hat{f}(\hat{x}, \hat{y}, \hat{z})\right)$ are plotted in Fig. 4.7. According to Fig. 4.7, the "○" marked line is used to indicate the solution by using parametric fuzzy arithmetic and the "□" marked line is used for the solution by fuzzy-affine arithmetic. The efficacy of fuzzy-affine arithmetic can be observed in the figure. In fuzzy-affine arithmetic, the solution bounds are comparatively tighter as compared to standard parametric fuzzy arithmetic.

For $a = 0$, the outer enclosure of the solution from parametric fuzzy arithmetic is found as $\tilde{f} = [-63, -11]$ and from fuzzy-affine arithmetic the outer enclosure of the solution is $\hat{f} = [-57, -17]$. Therefore for Example 4.16 , we may have

$$\hat{f} \subseteq \tilde{f}.$$

Example 4.17 Consider the fuzzy nonlinear function $f(x, y) = 2x^2 + 5y^2 - 3xy - x$, which is given in both Example 4.8 and Example 4.15. But in this example, we are assuming that

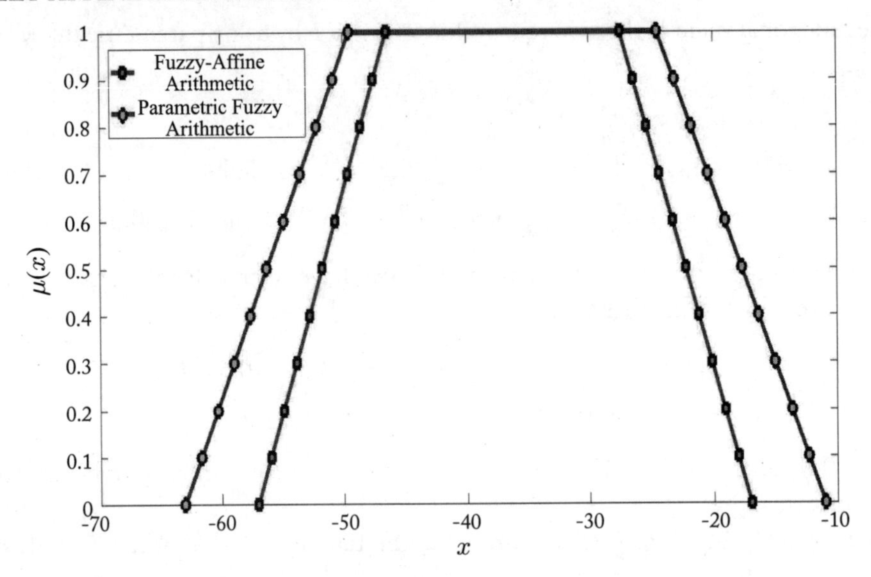

Figure 4.7: Comparison plot between fuzzy-affine arithmetic and parametric fuzzy arithmetic of Example 4.16.

all the consistent variables of the fuzzy nonlinear function are in the form of TFNs. That is, $\forall x \in \tilde{x} = (1, 2.5, 4)$ and $\forall y \in \tilde{y} = (4, 5.5, 7)$. Then, show that the fuzzy-affine arithmetic is more efficient and may give a tighter enclosure than the parametric fuzzy arithmetic by plotting both the fuzzy solutions.

Solution: To perform parametric fuzzy arithmetic, first we have to parameterize the given TFNs $\tilde{x} = (1, 2.5, 4)$ and $\tilde{y} = (4, 5.5, 7)$ into its interval parametric presentations by using a-cut technique as mention in Section 4.2.1. Thus,

$$\tilde{x}(a) = [1 + 1.5a, 4 - 1.5a] \quad \text{and} \quad \tilde{y}(a) = [4 + 1.5a, 7 - 1.5a], \quad \text{for} \quad a \in [0, 1].$$

Now, the operations of parametric fuzzy arithmetic (from Section 4.4) are applied to compute the functional value of the given fuzzy nonlinear function $\tilde{f}(\tilde{x}, \tilde{y})$ in interval parametric form. Hence, we may have

$$\begin{aligned}
\tilde{f}(\tilde{x}, \tilde{y}) &= 2\tilde{x}^2 + 5\tilde{y}^2 - 3\tilde{x}\tilde{y} - \tilde{x} \\
&= 2[1 + 1.5a, 4 - 1.5a]^2 + 5[4 + 1.5a, 7 - 1.5a]^2 \\
&\quad - 3[1 + 1.5a, 4 - 1.5a][4 + 1.5a, 7 - 1.5a] - [1 + 1.5a, 4 - 1.5a] \\
&\Rightarrow \tilde{f}(\tilde{x}(a), \tilde{y}(a)) = \left[9a^2 + 117a - 6, 9a^2 - 153a + 264\right], \quad \text{for} \quad a \in [0, 1].
\end{aligned}$$

Similarly, to perform fuzzy-affine arithmetic, we have to transform the consisting variables $\tilde{x}$ and $\tilde{y}$ into respective fuzzy-affine form (as given in Section 4.5.1) with different noise symbols

ε_1 and ε_2. Thus, the fuzzy-affine forms are found as follows:

$$\hat{x}\left(a, \varepsilon_1\right) = 2.5 + (1.5 - 1.5a)\varepsilon_1 \quad \text{and} \quad \hat{y}\left(a, \varepsilon_2\right) = 5.5 + (1.5 - 1.5a)\varepsilon_2,$$

where $a \in [0, 1]$ and $\varepsilon_1, \varepsilon_2 \in [-1, 1]$.

Further, by adopting fuzzy-affine arithmetic, the required functional value of the given nonlinear function $\hat{f}\left(\hat{x}, \hat{y}\right)$ in its fuzzy-affine form may be evaluated as

$$\begin{aligned}\hat{f}\left(\hat{x}, \hat{y}\right) &= 2\hat{x}^2 + 5\hat{y}^2 - 3\hat{x}\hat{y} - \hat{x} \\ &= 2\left(2.5 + (1.5 - 1.5a)\varepsilon_1\right)^2 + 5\left(5.5 + (1.5 - 1.5a)\varepsilon_2\right)^2 \\ &\quad - 3\left(2.5 + (1.5 - 1.5a)\varepsilon_1\right)\left(5.5 + (1.5 - 1.5a)\varepsilon_2\right) - \left(2.5 + (1.5 - 1.5a)\varepsilon_1\right) \\ &\Rightarrow \hat{f}\left(\hat{x}\left(a, \varepsilon_1\right), \hat{y}\left(a, \varepsilon_2\right)\right) \\ &= 120 - 7.5(1.5 - 1.5a)\varepsilon_1 + 47.5(1.5 - 1.5a)\varepsilon_2 \\ &\quad + 2(1.5 - 1.5a)^2\varepsilon_3 + 5(1.5 - 1.5a)^2\varepsilon_4 - 3(1.5 - 1.5a)^2\varepsilon_5\end{aligned}$$

where $a \in [0, 1]$ and $\varepsilon_i \in [-1, 1]$ for $i = 1, \ldots, 5$. Here, $\varepsilon_3 = (\varepsilon_1)^2$, $\varepsilon_4 = (\varepsilon_2)^2$ and $\varepsilon_5 = \varepsilon_1\varepsilon_2$ are the noise symbols newly generated during the operations of fuzzy-affine arithmetic.

It may be noted that, for $a = 0$, the outer enclosures of the functional value of the fuzzy nonlinear function by using parametric fuzzy arithmetic and fuzzy-affine arithmetic are $[-6, 264]$ and $[51, 189]$, respectively. Therefore, we may clearly have

$$f(x, y) \in \hat{f}\left(\hat{x}, \hat{y}\right) \subset \tilde{f}\left(\tilde{x}, \tilde{y}\right).$$

Now, let us plot a graph (See Fig. 4.8) of the corresponding TFN solutions computed by using both parametric fuzzy arithmetic $\left(\tilde{f}\left(\tilde{x}, \tilde{y}\right)\right)$ and fuzzy-affine arithmetic $\left(\hat{f}\left(\hat{x}, \hat{y}\right)\right)$. In Fig. 4.8, the line marked with "○" is used to indicate the solution by using parametric fuzzy arithmetic and the line marked with "□" is used to depict the solution by fuzzy-affine arithmetic.

Further, the TFN functional values of the given fuzzy nonlinear function $\tilde{f}\left(\tilde{x}, \tilde{y}\right)$ by using parametric fuzzy arithmetic as well as fuzzy-affine arithmetic for different values of the parameter a (that is for $a = 0, 0.2, 0.3, 0.5, 0.8, 0.9, 1$) are included in Table 4.1.

Therefore, it may be observed from the plots given in Fig. 4.8 and from Table 4.1 that the fuzzy-affine arithmetic is more efficient and results in tighter enclosures than the standard fuzzy arithmetic for the considered example (that is Example 4.17).

4.8 EXERCISES

4.1. Consider a TrFN $\tilde{T}r = (-50, -25, 0, 25)$. Convert the given TrFN into its interval parametric form. Further, multiply and divide the TrFN by itself by using the parametric fuzzy arithmetic.

4.2. Transform a TFN $\tilde{T} = (-111, -11, 89)$ into its fuzzy-affine form. Then, by adopting fuzzy-affine arithmetic, evaluate the square and cube of the given TFN.

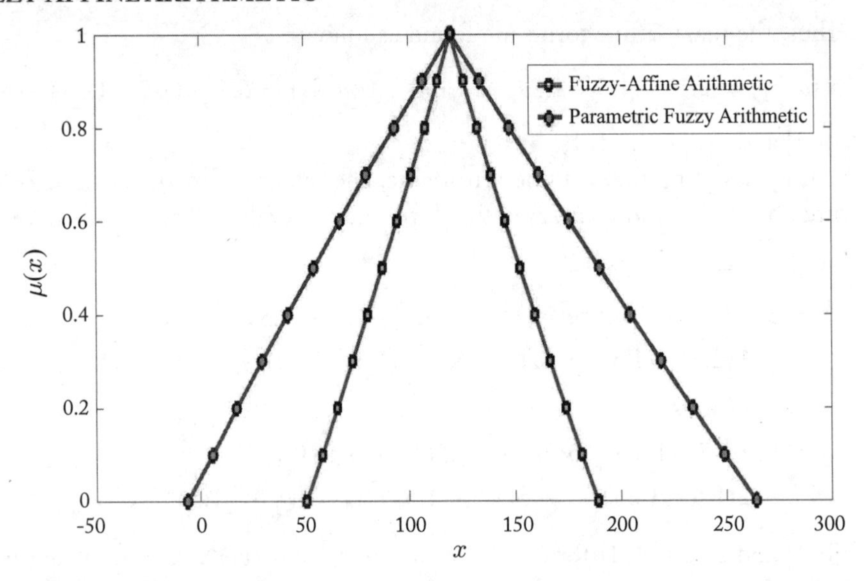

Figure 4.8: Comparison plot between fuzzy-affine arithmetic and parametric fuzzy arithmetic of Example 4.17.

Table 4.1: Comparison table between fuzzy-affine arithmetic and parametric fuzzy arithmetic of Example 4.17 for different values of a

a	Fuzzy-Affine Arithmetic		Parametric Fuzzy Arithmetic	
	Lower Bound ($\underline{f}$)	Upper Bound ($\overline{f}$)	Lower Bound ($\underline{f}$)	Upper Bound ($\overline{f}$)
0	51.00	189.00	-6.00	264.00
0.2	66.24	173.76	17.76	233.76
0.3	73.59	166.41	29.91	218.91
0.5	87.75	152.25	54.75	189.75
0.8	107.64	132.36	93.36	147.36
0.9	113.91	126.09	106.59	133.59
1	120.00	120.00	120.00	120.00

4.3. By using both parametric fuzzy arithmetic and fuzzy-affine arithmetic, evaluate the value of $\tilde{x}\tilde{y} - \tilde{y}\tilde{x}$, where the variables are in the form of TrFNs such that $\tilde{x} = (-5, 0, 2, 15)$ and $\tilde{y} = (-13, -9, -2, 5)$.

4.4. Find the fuzzy-affine form representation of the TFNs $\tilde{x} = (-5.7, -4, -3.65)$, $\tilde{y} = (1.5, 5.5, 9.5)$, and $\tilde{z} = (15, 17, 18)$. Thus, compute the fuzzy-affine solution of $15\tilde{x} - 3.6\tilde{y} + 10.5\tilde{z} - 9$ and further reconvert it to find its outer enclosure.

4.5. Let us consider a fuzzy nonlinear function such that $f(x, y) = x^2 + 2y^2 - 2xy$, where $\forall x \in \tilde{x} = (1, 2, 3)$ and $\forall y \in \tilde{y} = (5, 6, 7)$. Find the functional value of the fuzzy nonlinear function by utilizing both parametric fuzzy arithmetic and fuzzy-affine arithmetic. Further, plot the fuzzy solution and compare the results for both cases.

4.9 REFERENCES

[1] Chakraverty, S., Sahoo, D. M., and Mahato, N. R., 2019. *Concepts of Soft Computing: Fuzzy and ANN with Programming*. Springer. DOI: 10.1007/978-981-13-7430-2.

[2] Chakraverty, S., Tapaswini, S., and Behera, D., 2016. *Fuzzy Arbitrary Order System: Fuzzy Fractional Differential Equations and Applications*. John Wiley & Sons. DOI: 10.1002/9781119004233.

[3] Chakraverty, S., Tapaswini, S., and Behera, D., 2016. *Fuzzy Differential Equations and Applications for Engineers and Scientists*. CRC Press. DOI: 10.1201/9781315372853.

[4] Comba, J. L. D. and Stol, J., 1993. Affine arithmetic and its applications to computer graphics. In *Proc. of VI SIBGRAPI (Brazilian Symposium on Computer Graphics and Image Processing)*, pages 9–18.

[5] De Figueiredo, L. H. and Stolfi, J., 2004. Affine arithmetic: Concepts and applications. *Numerical Algorithms*, 37(1–4):147–158. DOI: 10.1023/b:numa.0000049462.70970.b6.

[6] Miyajima, S. and Kashiwagi, M., 2004. A dividing method utilizing the best multiplication in affine arithmetic. *IEICE Electronics Express*, 1(7):176–181. DOI: 10.1587/elex.1.176.

[7] Rout, S. and Chakraverty, S., 2019. Solving fully nonlinear eigenvalue problems of damped spring-mass structural systems using novel fuzzy-affine approach. *Computer Modeling in Engineering and Sciences*, 121(3):947–980. DOI: 10.32604/cmes.2019.08036.

[8] Skalna, I., 2009. Direct method for solving parametric interval linear systems with non-affine dependencies. In *International Conference on Parallel Processing and Applied Mathematics*, pages 485–494, Springer, Berlin, Heidelberg. DOI: 10.1007/978-3-642-14403-5_51.

[9] Skalna, I. and Hladík, M., 2017. A new algorithm for Chebyshev minimum-error multiplication of reduced affine forms. *Numerical Algorithms*, 76(4):1131–1152. DOI: 10.1007/s11075-017-0300-6.

[10] Stolfi, J. and De Figueiredo, L. H., 2003. An introduction to affine arithmetic. *Trends in Applied and Computational Mathematics*, 4(3):297–312. DOI: 10.5540/tema.2003.04.03.0297.

[11] Xu, C., Gu, W., Gao, F., Song, X., Meng, X., and Fan, M., 2016. Improved affine arithmetic based optimisation model for interval power flow analysis. *IET Generation, Transmission and Distribution*, 10(15):3910–3918. DOI: 10.1049/iet-gtd.2016.0601.

[12] Zadeh, L. A., 1965. Fuzzy sets. *Information and Control*, 8(3):338–353. DOI: 10.1016/s0019-9958(65)90241-x.

[13] Zadeh, L. A., Fu, K. S., and Tanaka, K., (Eds.), 2014. Fuzzy sets and their applications to cognitive and decision processes. *Proc. of the U.S.–Japan Seminar on Fuzzy Sets and their Applications*, Academic Press, University of California, Berkeley, CA, July 1–4, 1974. DOI: 10.1016/c2013-0-11734-5.

[14] Zimmermann, H. J., 2011. *Fuzzy Set Theory and its Applications.* Springer Science and Business Media. DOI: 10.1007/978-94-015-8702-0.

CHAPTER 5

Uncertain Static Problems

Under static conditions, the governing differential equations of various science and engineering problems lead to systems of simultaneous equations (linear and nonlinear). In this chapter, we focus on the solution for a system of linear equations. In mathematics, the theory of linear systems is the basis, and a fundamental part, of linear algebra. Also, the computational algorithms for finding the solutions for the system of linear equations are an important part of numerical linear algebra and play a prominent role in various fields viz. engineering, physics, chemistry, computer science and economics. For simplicity and easy computation, all the involved parameters and variables of the linear system are usually considered as deterministic or exact. But as a practical matter, due to the uncertain environment, one may have imprecise, incomplete, insufficient, or vague information about the parameters because of several errors. Traditionally, such uncertainty or vagueness may be modeled through a probabilistic approach. But a large amount of data is required for the traditional probabilistic approach. Without a sufficient amount of experimental data, the probabilistic methods may not deliver reliable results at the required precision. Therefore, intervals and/or fuzzy numbers may become efficient tools to handle uncertain and vague parameters when there is an insufficient amount of data available. In this regard, uncertain static problems may be modeled through an interval system of linear equations (ISLE) ($[A][x] = [b]$) and/or fuzzy system of linear equations (FSLE) ($\tilde{A}\tilde{x} = \tilde{b}$).

A system of linear equations has a wide variety of application problems in several fields of engineering and sciences. For instance, under static conditions, the equation of motion for the structural vibrational problems $M\ddot{s}(t) + C\dot{s}(t) + Ks(t) = f(t)$ may lead to a system of linear equations $Mx = f$, where M is the mass matrix and f is the external load vector. By solving this linear system, we may evaluate the nodal displacement vector (x) for the given system. Further, the static analysis of circuits leads to the linear system $RI = V$, where R is the resistance matrix, V is the voltage vector, and I is the unknown vector for current flow. Therefore, the solution for linear systems plays a major role in various application problems.

5.1 CRISP SYSTEM OF LINEAR EQUATIONS (CSLE)

The system of linear equations having crisp (exact) parameters may be denoted as

$$Ax = b, \tag{5.1}$$

where A is known as the coefficient matrix of order $p \times p$, x is the $p \times 1$ solution vector, and b is the $p \times 1$ column vector.

There exist many well-known methods for the solution of systems of linear equations viz. row reduction (Gaussian elimination), Cramer's rule, matrix inverse method ($x = A^{-1} * b$), LU decomposition, Cholesky decomposition, Levinson recursion method, quantum algorithm, and many more. Further, there also exist different numerical methods to handle systems of linear equations such as the Gauss–Jacobi method, Gauss–Seidel method, SOR algorithm method, multigrid method, and conjugate gradient method.

Example 5.1 Compute the solution vector of the linear system $Ax = b$, where

$$A = \begin{pmatrix} 1 & 1 & -2 & 1 & 3 & -1 \\ 2 & 1 & 1 & 2 & 1 & -3 \\ 1 & 3 & -3 & -1 & 2 & 1 \\ 5 & 2 & -1 & -1 & 2 & 1 \\ -3 & 1 & 2 & 3 & 1 & 3 \\ 4 & 3 & 1 & -6 & -3 & -2 \end{pmatrix} \quad \text{and} \quad b = \begin{pmatrix} 4 \\ 20 \\ -15 \\ -3 \\ 16 \\ -27 \end{pmatrix}.$$

Solution: Here the above system is of order 6×6. The solution vector of the given system may easily be computed as

$$x = \begin{pmatrix} 1 & -2 & 3 & 4 & 2 & -1 \end{pmatrix}^T.$$

5.2 INTERVAL SYSTEM OF LINEAR EQUATIONS (ISLE)

When the uncertain parameters are considered as closed intervals, the uncertain static problem may be addressed as an interval system of linear equations (ISLE). Then the ISLE may be denoted as

$$[A][x] = [b], \tag{5.2}$$

where $[A]$ is the $p \times p$ interval coefficient matrix (whose entries are taken in the form of closed intervals), $[x]$ is the $p \times 1$ interval solution vector containing interval unknowns, and $[b]$ is the $p \times 1$ interval column vector.

Suppose $[a_{mn}]$ (for $1 \leq m, n \leq p$) denotes the elements (in form of closed intervals) of the interval coefficient matrix $[A]$. Similarly, $[x_n]$ and $[b_m]$ denote the elements of $[x]$ and $[b]$, respectively. Thus, Eq. (5.2) may further be written as

$$\sum_{n=1}^{p} [a_{mn}]\,[x_n] = [b_m], \quad \text{for} \quad m = 1, \ldots, p. \tag{5.3}$$

Expanding the above system given in Eq. (5.3), we may have

$$\begin{aligned}
[a_{11}][x_1] + [a_{12}][x_2] + \cdots + [a_{1p}][x_p] &= [b_1] \\
[a_{21}][x_1] + [a_{22}][x_2] + \cdots + [a_{2p}][x_p] &= [b_2] \\
&\vdots \\
[a_{p1}][x_1] + [a_{p2}][x_2] + \cdots + [a_{pp}][x_p] &= [b_p].
\end{aligned} \tag{5.4}$$

Suppose the lower and upper bounds of all the consisting elements of the system (5.4) are in the form given below.

$$[a_{mn}] = \left[\underline{a_{mn}}, \overline{a_{mn}}\right], \quad [x_n] = \left[\underline{x_n}, \overline{x_n}\right], \quad \text{and} \quad [b_m] = \left[\underline{b_m}, \overline{b_m}\right].$$

Thus, the system (5.4) in terms of its lower and upper bounds may be expressed as follows:

$$\begin{aligned}
\left[\underline{a_{11}}, \overline{a_{11}}\right]\left[\underline{x_1}, \overline{x_1}\right] + \left[\underline{a_{12}}, \overline{a_{12}}\right]\left[\underline{x_2}, \overline{x_2}\right] + \cdots + \left[\underline{a_{1p}}, \overline{a_{1p}}\right]\left[\underline{x_p}, \overline{x_p}\right] &= \left[\underline{b_1}, \overline{b_1}\right] \\
\left[\underline{a_{21}}, \overline{a_{21}}\right]\left[\underline{x_1}, \overline{x_1}\right] + \left[\underline{a_{22}}, \overline{a_{22}}\right]\left[\underline{x_2}, \overline{x_2}\right] + \cdots + \left[\underline{a_{2p}}, \overline{a_{2p}}\right]\left[\underline{x_p}, \overline{x_p}\right] &= \left[\underline{b_2}, \overline{b_2}\right] \\
&\vdots \\
\left[\underline{a_{p1}}, \overline{a_{p1}}\right]\left[\underline{x_1}, \overline{x_1}\right] + \left[\underline{a_{p2}}, \overline{a_{p2}}\right]\left[\underline{x_2}, \overline{x_2}\right] + \cdots + \left[\underline{a_{pp}}, \overline{a_{pp}}\right]\left[\underline{x_p}, \overline{x_p}\right] &= \left[\underline{b_p}, \overline{b_p}\right].
\end{aligned} \tag{5.5}$$

In matrix form, the above ISLE (5.5) may be written as

$$\begin{pmatrix}
\left[\underline{a_{11}}, \overline{a_{11}}\right] & \left[\underline{a_{12}}, \overline{a_{12}}\right] & \cdots & \left[\underline{a_{1p}}, \overline{a_{1p}}\right] \\
\left[\underline{a_{21}}, \overline{a_{21}}\right] & \left[\underline{a_{22}}, \overline{a_{22}}\right] & \cdots & \left[\underline{a_{2p}}, \overline{a_{2p}}\right] \\
\vdots & \vdots & \ddots & \vdots \\
\left[\underline{a_{p1}}, \overline{a_{p1}}\right] & \left[\underline{a_{p2}}, \overline{a_{p2}}\right] & \cdots & \left[\underline{a_{pp}}, \overline{a_{pp}}\right]
\end{pmatrix} \cdot \begin{pmatrix}
\left[\underline{x_1}, \overline{x_1}\right] \\
\left[\underline{x_2}, \overline{x_2}\right] \\
\vdots \\
\left[\underline{x_p}, \overline{x_p}\right]
\end{pmatrix} = \begin{pmatrix}
\left[\underline{b_1}, \overline{b_1}\right] \\
\left[\underline{b_2}, \overline{b_2}\right] \\
\vdots \\
\left[\underline{b_p}, \overline{b_p}\right]
\end{pmatrix}. \tag{5.6}$$

5.3 SOLUTION OF ISLE

For simplicity, let us first consider a 2×2 ISLE, and using the proposed method described below, the system can be solved by using affine arithmetic computations. The resulting solution is found in the form of affine representations which may be further converted to compute the interval bounds of the ISLE.

The 2×2 ISLE $[A][x] = [b]$ may be expressed as

$$\begin{pmatrix}
\left[\underline{a_{11}}, \overline{a_{11}}\right] & \left[\underline{a_{12}}, \overline{a_{12}}\right] \\
\left[\underline{a_{21}}, \overline{a_{21}}\right] & \left[\underline{a_{22}}, \overline{a_{22}}\right]
\end{pmatrix} \cdot \begin{pmatrix}
\left[\underline{x_1}, \overline{x_1}\right] \\
\left[\underline{x_2}, \overline{x_2}\right]
\end{pmatrix} = \begin{pmatrix}
\left[\underline{b_1}, \overline{b_1}\right] \\
\left[\underline{b_2}, \overline{b_2}\right]
\end{pmatrix}. \tag{5.7}$$

Suppose A_c and A_Δ are the center and radius matrices of interval coefficient matrix $[A]$ of the ISLE (5.7), respectively; they may be defined as

$$A_c = \begin{pmatrix} a_{11}^{(0)} & a_{12}^{(0)} \\ a_{21}^{(0)} & a_{22}^{(0)} \end{pmatrix} \quad \text{and} \quad A_\Delta = \begin{pmatrix} a_{11}^{(1)} & a_{12}^{(1)} \\ a_{21}^{(1)} & a_{22}^{(1)} \end{pmatrix}, \tag{5.8}$$

where $a_{mn}^{(0)} = \frac{1}{2}\left(\underline{a_{mn}} + \overline{a_{mn}}\right)$ and $a_{mn}^{(1)} = \frac{1}{2}\left(\overline{a_{mn}} - \underline{a_{mn}}\right)$ for $m, n = 1, 2$.

Thus, the interval coefficient matrix $[A]$ may be converted into its affine form representation denoted by $\hat{A}$ as

$$\hat{A} = \begin{pmatrix} a_{11}^{(0)} + a_{11}^{(1)}\varepsilon_1 & a_{12}^{(0)} + a_{12}^{(1)}\varepsilon_2 \\ a_{21}^{(0)} + a_{21}^{(1)}\varepsilon_3 & a_{22}^{(0)} + a_{22}^{(1)}\varepsilon_4 \end{pmatrix}, \tag{5.9}$$

where $\varepsilon_i \in [-1, 1]$ for $i = 1, 2, 3, 4$ are noise symbols.

Similarly, the respective affine form representations of the interval vectors $[x]$ and $[b]$ are

$$\hat{x} = \begin{pmatrix} x_1^{(0)} + x_1^{(1)}\varepsilon_5 \\ x_2^{(0)} + x_2^{(1)}\varepsilon_6 \end{pmatrix} \quad \text{and} \quad \hat{b} = \begin{pmatrix} b_1^{(0)} + b_1^{(1)}\varepsilon_7 \\ b_2^{(0)} + b_2^{(1)}\varepsilon_8 \end{pmatrix},$$

where $\varepsilon_i \in [-1, 1]$ for $i = 5, 6, 7, 8$ are noise symbols.

Therefore, the ISLE (given in Eq. (5.7)) is converted into its affine form representation known as the affine system of linear equations (ASLE), denoted as

$$\hat{A}\hat{x} = \hat{b}. \tag{5.10}$$

The above 2×2 ASLE may be expressed as follows:

$$\begin{pmatrix} a_{11}^{(0)} + a_{11}^{(1)}\varepsilon_1 & a_{12}^{(0)} + a_{12}^{(1)}\varepsilon_2 \\ a_{21}^{(0)} + a_{21}^{(1)}\varepsilon_3 & a_{22}^{(0)} + a_{22}^{(1)}\varepsilon_4 \end{pmatrix} \cdot \begin{pmatrix} x_1^{(0)} + x_1^{(1)}\varepsilon_5 \\ x_2^{(0)} + x_2^{(1)}\varepsilon_6 \end{pmatrix} = \begin{pmatrix} b_1^{(0)} + b_1^{(1)}\varepsilon_7 \\ b_2^{(0)} + b_2^{(1)}\varepsilon_8 \end{pmatrix}. \tag{5.11}$$

The above system can be written as follows:

$$\left.\begin{aligned} &\left(a_{11}^{(0)} + a_{11}^{(1)}\varepsilon_1\right)\left(x_1^{(0)} + x_1^{(1)}\varepsilon_5\right) + \left(a_{12}^{(0)} + a_{12}^{(1)}\varepsilon_2\right) \\ &\qquad\qquad \left(x_2^{(0)} + x_2^{(1)}\varepsilon_6\right) = b_1^{(0)} + b_1^{(1)}\varepsilon_7 \\ &\left(a_{21}^{(0)} + a_{21}^{(1)}\varepsilon_3\right)\left(x_1^{(0)} + x_1^{(1)}\varepsilon_5\right) + \left(a_{22}^{(0)} + a_{22}^{(1)}\varepsilon_4\right) \\ &\qquad\qquad \left(x_2^{(0)} + x_2^{(1)}\varepsilon_6\right) = b_2^{(0)} + b_2^{(1)}\varepsilon_8 \end{aligned}\right\}. \tag{5.12}$$

From Eq. (5.12), we may have the following particular form of 2×2 linear system with $x_1^{(0)}$ and $x_2^{(0)}$ as unknown variables.

$$\left.\begin{aligned} a_{11}^{(0)}x_1^{(0)} + a_{12}^{(0)}x_2^{(0)} &= b_1^{(0)} \\ a_{21}^{(0)}x_1^{(0)} + a_{22}^{(0)}x_2^{(0)} &= b_2^{(0)} \end{aligned}\right\}. \tag{5.13}$$

The above linear system has crisp (exact) parameters and may be solved to compute the variables $x_1{}^{(0)}$ and $x_2{}^{(0)}$ by using various well-known approaches. Further, during affine multiplication, in the equations of the affine system (5.12), new noise symbols may also be generated. Then, comparing the total deviations from both sides of the affine linear system (5.12), the following 2×2 system having unknown variables $x_1{}^{(1)}$ and $x_2{}^{(1)}$ may be generated.

$$\left.\begin{array}{r} \left|a_{11}{}^{(0)}x_1{}^{(1)}\right| + \left|a_{11}{}^{(1)}x_1{}^{(0)}\right| + \left|a_{11}{}^{(1)}x_1{}^{(1)}\right| + \left|a_{12}{}^{(0)}x_2{}^{(1)}\right| + \left|a_{12}{}^{(1)}x_2{}^{(0)}\right| \\ + \left|a_{12}{}^{(1)}x_2{}^{(1)}\right| = \left|b_1{}^{(1)}\right| \\ \\ \left|a_{21}{}^{(0)}x_1{}^{(1)}\right| + \left|a_{21}{}^{(1)}x_1{}^{(0)}\right| + \left|a_{21}{}^{(1)}x_1{}^{(1)}\right| + \left|a_{22}{}^{(0)}x_2{}^{(1)}\right| + \left|a_{22}{}^{(1)}x_2{}^{(0)}\right| \\ + \left|a_{22}{}^{(1)}x_2{}^{(1)}\right| = \left|b_2{}^{(1)}\right| \end{array}\right\}. \tag{5.14}$$

Thus, plugging the solutions of the system (5.13) in the above system (5.14) and solving by applying any known approach, the variables $x_1{}^{(1)}$ and $x_2{}^{(1)}$ are also evaluated. Hence, the solution of the above affine system of equations (5.11) by using the proposed procedure can be written as

$$\hat{x} = \begin{pmatrix} x_1{}^{(0)} + x_1{}^{(1)}\varepsilon_5 \\ x_2{}^{(0)} + x_2{}^{(1)}\varepsilon_6 \end{pmatrix}. \tag{5.15}$$

By converting the affine vector solution (5.15) into interval, we may have the solution vector of the interval system (5.7) as

$$\begin{pmatrix} [\underline{x_1}, \overline{x_1}] \\ [\underline{x_2}, \overline{x_2}] \end{pmatrix}, \quad \text{where} \quad \underline{x_n} = x_n{}^{(0)} - x_n{}^{(1)} \quad \text{and} \quad \overline{x_n} = x_n{}^{(0)} + x_n{}^{(1)} \quad \text{for } n = 1, 2. \tag{5.16}$$

Example 5.2 Find the affine solution vector of the 2×2 ISLE $[A][x] = [b]$ [Chakraverty et. al. (2019) [9]], where

$$[A] = \begin{pmatrix} [9, 11] & [-4.1, -3.9] \\ [-4.1, -3.9] & [15, 16] \end{pmatrix} \quad \text{and} \quad [b] = \begin{pmatrix} [26, 30] \\ [35, 39] \end{pmatrix}.$$

and then convert it to compute its interval bounds.

Solution: The center and radius of the given interval coefficient matrix $[A]$ and the interval column vector $[b]$ is

$$A_c = \begin{pmatrix} 10 & -4 \\ -4 & 15.5 \end{pmatrix}; \quad A_\Delta = \begin{pmatrix} 1 & 0.1 \\ 0.1 & 0.5 \end{pmatrix} \quad \text{and}$$

$$b_c = \begin{pmatrix} 28 \\ 37 \end{pmatrix}; \quad b_\Delta = \begin{pmatrix} 2 \\ 2 \end{pmatrix}.$$

Table 5.1: Interval bounds of the solution vector of ISLE for Example 5.2

i	Present Method		Chakraverty et al. (2019)	
	Lower Bound ($\underline{x_i}$)	Upper Bound ($\overline{x_i}$)	Lower Bound ($\underline{x_i}$)	Upper Bound ($\overline{x_i}$)
1	3.9364	4.4378	3.9312	4.5139
2	3.4129	3.5223	3.3957	3.5671

Thus, the given ISLE is converted to the affine system of linear equations $\hat{A}\hat{x} = \hat{b}$ (given in Eq. (5.11)) as follows:

$$\begin{pmatrix} 10+\varepsilon_1 & -4+0.1\varepsilon_2 \\ -4+0.1\varepsilon_3 & 15.5+0.5\varepsilon_4 \end{pmatrix} \cdot \begin{pmatrix} {x_1}^{(0)} + {x_1}^{(1)}\varepsilon_5 \\ {x_2}^{(0)} + {x_2}^{(1)}\varepsilon_6 \end{pmatrix} = \begin{pmatrix} 28+2\varepsilon_7 \\ 37+2\varepsilon_8 \end{pmatrix}. \tag{5.17}$$

Now, by performing affine multiplication to the above affine system similarly to Eq. (5.13), we may have the following 2×2 crisp linear system (5.18).

$$\begin{aligned} 10{x_1}^{(0)} - 4{x_2}^{(0)} &= 28 \\ -4{x_1}^{(0)} + 15.5{x_2}^{(0)} &= 37. \end{aligned} \tag{5.18}$$

Further, comparing the total deviations from both sides of the affine system (5.17), we may have the below (5.19) system

$$\begin{aligned} 11{x_1}^{(1)} + {x_1}^{(0)} + 4.1{x_2}^{(1)} + 0.1{x_2}^{(0)} &= 2 \\ 4.1{x_1}^{(1)} + 0.1{x_1}^{(0)} + 16{x_2}^{(1)} + 0.5{x_2}^{(0)} &= 2. \end{aligned} \tag{5.19}$$

Hence, solving both the systems (5.18) and (5.19) all the unknown variables ${x_1}^{(0)}$, ${x_2}^{(0)}$, ${x_1}^{(1)}$, and ${x_2}^{(1)}$ are computed. Therefore, the affine solution vector of the given ISLE is calculated as

$$\hat{x} = \begin{pmatrix} 4.1871 - 0.2507\varepsilon_5 \\ 3.4676 + 0.0547\varepsilon_6 \end{pmatrix}. \tag{5.20}$$

Converting the affine solution vector (5.20) into its interval form, adopting Section 3.3, the interval bounds of the interval solution vector of the given ISLE are calculated. The values of interval bounds are illustrated in Table 5.1. A comparison of the solution bounds with the solution of Chakraverty et. al. (2019) [9] is also included in the table. From Table 5.1, it is clearly evident that the present method works efficiently and results obtained are with tighter enclosures.

In general, let us consider a $p \times p$ ISLE $[A][x] = [b]$ as given in Eq. (5.3) as

$$\sum_{n=1}^{p} \left[\underline{a_{mn}}, \overline{a_{mn}}\right]\left[\underline{x_n}, \overline{x_n}\right] = \left[\underline{b_m}, \overline{b_m}\right], \quad \text{for} \quad m = 1, \ldots, p. \tag{5.21}$$

The matrix form of the above system (5.21) may be expressed as

$$\begin{pmatrix} [\underline{a_{11}}, \overline{a_{11}}] & [\underline{a_{12}}, \overline{a_{12}}] & \cdots & [\underline{a_{1p}}, \overline{a_{1p}}] \\ [\underline{a_{21}}, \overline{a_{21}}] & [\underline{a_{22}}, \overline{a_{22}}] & \cdots & [\underline{a_{2p}}, \overline{a_{2p}}] \\ \vdots & \vdots & \ddots & \\ [\underline{a_{p1}}, \overline{a_{p1}}] & [\underline{a_{p2}}, \overline{a_{p2}}] & \cdots & [\underline{a_{pp}}, \overline{a_{pp}}] \end{pmatrix} \cdot \begin{pmatrix} [\underline{x_1}, \overline{x_1}] \\ [\underline{x_2}, \overline{x_2}] \\ \vdots \\ [\underline{x_p}, \overline{x_p}] \end{pmatrix} = \begin{pmatrix} [\underline{b_1}, \overline{b_1}] \\ [\underline{b_2}, \overline{b_2}] \\ \vdots \\ [\underline{b_p}, \overline{b_p}] \end{pmatrix}. \tag{5.22}$$

Thus, the respective center and radius matrices of the $p \times p$ coefficient matrix and $p \times 1$ solution column vector and $p \times 1$ right-hand side column vector of the above system (5.22) may be written as

$$A_c = \begin{pmatrix} a_{11}{}^{(0)} & a_{12}{}^{(0)} & \cdots & a_{1p}{}^{(0)} \\ a_{21}{}^{(0)} & a_{22}{}^{(0)} & \cdots & a_{2p}{}^{(0)} \\ \vdots & \vdots & \ddots & \vdots \\ a_{p1}{}^{(0)} & a_{p2}{}^{(0)} & \cdots & a_{pp}{}^{(0)} \end{pmatrix} \quad \text{and} \quad A_\Delta = \begin{pmatrix} a_{11}{}^{(1)} & a_{12}{}^{(1)} & \cdots & a_{1p}{}^{(1)} \\ a_{21}{}^{(1)} & a_{22}{}^{(1)} & \cdots & a_{2p}{}^{(1)} \\ \vdots & \vdots & \ddots & \vdots \\ a_{p1}{}^{(1)} & a_{p2}{}^{(1)} & \cdots & a_{pp}{}^{(1)} \end{pmatrix};$$

$$x_c = \begin{pmatrix} x_1{}^{(0)} \\ x_2{}^{(0)} \\ \vdots \\ x_p{}^{(0)} \end{pmatrix} \quad \text{and} \quad x_\Delta = \begin{pmatrix} x_1{}^{(1)} \\ x_2{}^{(1)} \\ \vdots \\ x_p{}^{(1)} \end{pmatrix}; \quad b_c = \begin{pmatrix} b_1{}^{(0)} \\ b_2{}^{(0)} \\ \vdots \\ b_p{}^{(0)} \end{pmatrix} \quad \text{and} \quad x_\Delta = \begin{pmatrix} b_1{}^{(0)} \\ b_2{}^{(0)} \\ \vdots \\ b_p{}^{(0)} \end{pmatrix}.$$

Here, $a_{mn}{}^{(0)} = \frac{1}{2}\left(\underline{a_{mn}} + \overline{a_{mn}}\right)$ and $a^{(1)} = \frac{1}{2}\left(\overline{a_{mn}} - \underline{a_{mn}}\right)$; $x_n{}^{(0)} = \frac{1}{2}\left(\underline{x_n} + \overline{x_n}\right)$ and $x_n{}^{(1)} = \frac{1}{2}\left(\overline{x_n} - \underline{x_n}\right)$; $b_m{}^{(0)} = \frac{1}{2}\left(\underline{b_m} + \overline{b_m}\right)$ and $a_m{}^{(1)} = \frac{1}{2}\left(\overline{b_m} - \underline{b_m}\right)$, for $m, n = 1, 2, \ldots, p$.

Now, by using the above center and radius matrices, all the consisting matrices of the system (5.22) can be converted into its affine form representations. Suppose $\hat{A}\hat{x} = \hat{b}$ denotes the $p \times p$ affine linear system of equations, then its matrix form may be expressed as

$$\begin{pmatrix} a_{11}{}^{(0)} + a_{11}{}^{(1)}\varepsilon_{11} & a_{12}{}^{(0)} + a_{12}{}^{(1)}\varepsilon_{12} & & a_{1p}{}^{(0)} + a_{1p}{}^{(1)}\varepsilon_{1p} \\ a_{21}{}^{(0)} + a_{21}{}^{(1)}\varepsilon_{21} & a_{22}{}^{(0)} + a_{22}{}^{(1)}\varepsilon_{22} & & a_{2p}{}^{(0)} + a_{2p}{}^{(1)}\varepsilon_{2p} \\ \vdots & \vdots & \ddots & \vdots \\ a_{p1}{}^{(0)} + a_{p1}{}^{(1)}\varepsilon_{p1} & a_{p2}{}^{(0)} + a_{p2}{}^{(1)}\varepsilon_{p2} & & a_{pp}{}^{(0)} + a_{pp}{}^{(1)}\varepsilon_{pp} \end{pmatrix} \cdot \begin{pmatrix} x_1{}^{(0)} + x_1{}^{(1)}\varepsilon_{x1} \\ x_2{}^{(0)} + x_1{}^{(1)}\varepsilon_{x2} \\ \vdots \\ x_p{}^{(0)} + x_p{}^{(1)}\varepsilon_{xp} \end{pmatrix}$$

$$= \begin{pmatrix} b_1{}^{(0)} + b_1{}^{(1)}\varepsilon_{b1} \\ b_2{}^{(0)} + b_2{}^{(1)}\varepsilon_{b2} \\ \vdots \\ b_p{}^{(0)} + b_p{}^{(1)}\varepsilon_{bp} \end{pmatrix}, \tag{5.23}$$

where $\varepsilon_{mn}, \varepsilon_{xn}, \varepsilon_{bm} \in \mathbb{D} = [-1, 1]$ for $m, n = 1, \ldots, p$ are noise symbols of different affine numbers present in the system (5.23).

Further, performing affine multiplication in the affine system (5.23), new noise symbols will be generated. Then from the affine system (5.23), the following two $p \times p$ systems of linear equations may be obtained (as mentioned in Eqs. (5.13) and (5.14)).

$$\sum_{n=1}^{p} a_{mn}{}^{(0)} x_n{}^{(0)} = b_m{}^{(0)}, \quad \text{for} \quad m = 1, 2, \ldots, p. \tag{5.24}$$

$$\sum_{n=1}^{p} \left[\left| a_{mn}{}^{(0)} x_n{}^{(1)} \right| + \left| a_{mn}{}^{(1)} x_n{}^{(0)} \right| + \left| a_{mn}{}^{(1)} x_n{}^{(1)} \right| \right] = \left| b_m{}^{(1)} \right|, \quad \text{for} \quad m = 1, 2, \ldots, p. \tag{5.25}$$

Solving the above two systems (5.24) and (5.25) by adopting any well-known methods, the consisting unknown variables $x_1{}^{(0)}, x_2{}^{(0)}, \ldots, x_p{}^{(0)}$ (of the system (5.24)) and $x_1{}^{(1)}, x_2{}^{(1)}, \ldots, x_p{}^{(1)}$ (of the system (5.25)) may be evaluated. Hence, the affine solution vector of the affine system (5.23) is found as

$$\hat{x} = \begin{pmatrix} x_1{}^{(0)} + x_1{}^{(1)} \varepsilon_{x1} \\ x_2{}^{(0)} + x_2{}^{(1)} \varepsilon_{x2} \\ \\ x_p{}^{(0)} + x_p{}^{(1)} \varepsilon_{xp} \end{pmatrix}, \tag{5.26}$$

where $\varepsilon_{xn} \in \mathbb{D} = [-1, 1]$ for $n = 1, 2, \ldots, p$ are noise symbols of the elements of the affine solution vector that may not present in other affine form representations. Finally, the affine solution vector (5.26) is transformed into its interval bounds (as given in Section 3.3), and the interval solution vector (5.27) of the ISLE (5.22) may be estimated.

$$[x] = \begin{pmatrix} \left[\underline{x_1}, \overline{x_1}\right] \\ \left[\underline{x_2}, \overline{x_2}\right] \\ \vdots \\ \left[\underline{x_p}, \overline{x_p}\right] \end{pmatrix},$$

$$\text{where} \quad \underline{x_n} = x_n{}^{(0)} - x_n{}^{(1)} \quad \text{and} \quad \overline{x_n} = x_n{}^{(0)} + x_n{}^{(1)} \quad \text{for} \quad n = 1, 2, \ldots, p. \tag{5.27}$$

Example 5.3 Let us consider now an application problem of structures having uncertainties in terms of closed intervals. For this example, a three-bar truss as depicted in Fig. 5.1 [Bhavikatti (2005) [6]] is used for investigation. In this particular case, only the horizontal displacement ($\tilde{s}_n$ for $n = 1, 3, 5$) and vertical displacement ($\tilde{s}_n$ for $n = 2, 4, 6$) of the nodes are considered. Here, the truss is subject to the external load acting on node 3. All the consisting material properties

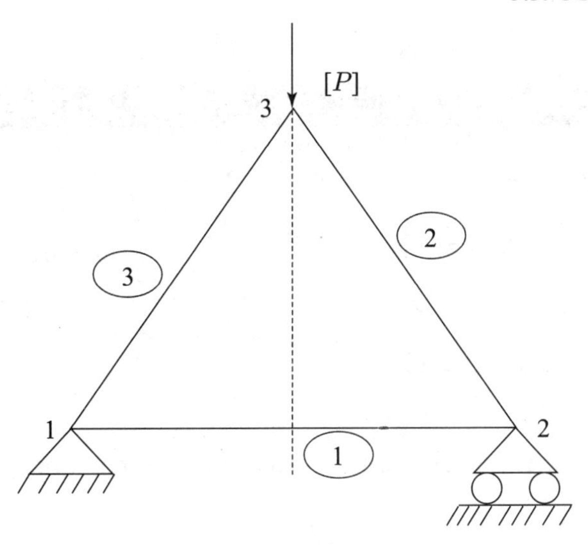

Figure 5.1: Three-bar truss.

Table 5.2: Material properties of the three-bar truss of Example 5.3

Parameters	Values
E_n for $n = 1, 2, 3$ (kN/m^2)	200
l_1 (mm)	800
l_n for $n = 2, 3$ (mm)	$400\sqrt{2}$
A_1 (mm^2)	1,500
A_n for $n = 2, 3$ (mm^2)	2,000
$[P]$ (kN)	$[-160, -140]$

viz. Young's modulus, length of three sections, cross-sectional area, and external load of the three-bar truss are listed in Table 5.2 [Bhavikatti (2005) [6]].

With these material and geometric properties, two cases are considered here. In Case I, the stiffness matrix is taken as a crisp (exact) matrix, but in Case II, the stiffness matrix is considered in terms of an interval matrix. Finally, for both the cases, the external load acting at node 3 are assumed to be uncertain and may be taken in terms of intervals.

In this regard, we may have two 3 × 3 ISLE for the two cases given as follows. By solving these systems with the proposed procedure given in Section 5.3, we may have the uncertain static responses corresponding to both the cases. It is worth mentioning that the crisp form of the model may be written as $Kx = f$, where K and f are the respective stiffness matrix and

Table 5.3: Affine solution vector of the ISLE for Example 5.3

Case I	Case II
$\hat{s} = \begin{pmatrix} 0.1981 - 0.0070\varepsilon_{10} \\ 0.0961 - 0.0064\varepsilon_{11} \\ -0.3121 + 0.0208\varepsilon_{12} \end{pmatrix}$	$\hat{s} = \begin{pmatrix} 0.1981 - 0.0072\varepsilon_{10} \\ 0.0961 - 0.0062\varepsilon_{11} \\ -0.3121 + 0.0211\varepsilon_{12} \end{pmatrix}$

external force vector. The two uncertain cases (from the respective mathematical model of the system) are taken here as

Case I.

$$\begin{pmatrix} 728.55 & -353.55 & 353.55 \\ 0 & 556.98 & 171.57 \\ 0 & 0 & 480.56 \end{pmatrix} \cdot \begin{pmatrix} [\underline{s_3}, \overline{s_3}] \\ [\underline{s_5}, \overline{s_5}] \\ [\underline{s_6}, \overline{s_6}] \end{pmatrix} = \begin{pmatrix} 0 \\ 0 \\ [-160, -140] \end{pmatrix}.$$

Case II.

$$\begin{pmatrix} [728.35, 728.75] & [-353.85, -353.25] & [353.25, 353.85] \\ 0 & [556.88, 557.08] & [171.07, 172.07] \\ 0 & 0 & [480.16, 480.96] \end{pmatrix} \cdot \begin{pmatrix} [\underline{s_3}, \overline{s_3}] \\ [\underline{s_5}, \overline{s_5}] \\ [\underline{s_6}, \overline{s_6}] \end{pmatrix}$$
$$= \begin{pmatrix} 0 \\ 0 \\ [-160, -140] \end{pmatrix}.$$

These systems are converted into their affine form representation and are solved (as given in Example 5.2) to find the affine solution vector listed in Table 5.3.

The above affine solutions may be converted into their interval bounds by using the procedure as mentioned earlier, and the lower and upper bounds of the interval solution vector of the given ISLE are incorporated in Table 5.4.

5.4 FUZZY SYSTEM OF LINEAR EQUATIONS (FSLE)

The uncertain static problem may be referred as the fuzzy system of linear equations (FSLE) if all the consisting uncertain parameters of the problem are in the form of fuzzy numbers viz.

Table 5.4: Interval bounds of the solution vector of ISLE for Example 5.3

	Case I		Case II	
i	Lower Bound ($\underline{v_i}$)	Upper Bound ($\overline{v_i}$)	Lower Bound ($\underline{v_i}$)	Upper Bound ($\overline{v_i}$)
3	0.1911	0.2051	0.1909	0.2053
5	0.0897	0.1025	0.0899	0.1023
6	-0.3329	-0.2913	-0.3332	-0.2910

triangular fuzzy number, trapezoidal fuzzy number, Gaussian fuzzy number, and so on. Then the FSLE may be presented as

$$\tilde{A}\tilde{x} = \tilde{b}, \tag{5.28}$$

where $\tilde{A}$ is known as the fuzzy coefficient matrix (whose entries are taken in the form of fuzzy numbers) having dimension $p \times p$, $\tilde{x}$ is the fuzzy solution vector, and $\tilde{b}$ is the fuzzy column vector of dimensions $p \times 1$ and $p \times 1$, respectively.

Let us suppose the elements of the fuzzy coefficient matrix $\tilde{A}$ are in the form of $\tilde{A} = (\tilde{a}_{mn})$ (for $1 \leq m, n \leq p$) and the elements of the other two vectors are $\tilde{x} = (\tilde{x}_n)$ and $\tilde{b} = (\tilde{b}_m)$. Thus, Eq. (5.28) may be expressed as

$$\sum_{n=1}^{p} \tilde{a}_{mn}\tilde{x}_n = \tilde{b}_m, \quad \text{for} \quad m = 1, \ldots, p. \tag{5.29}$$

Expanding the above system given in Eq. (5.29), we may have

$$\begin{aligned} \tilde{a}_{11}\tilde{x}_1 + \tilde{a}_{12}\tilde{x}_2 + \cdots + \tilde{a}_{1p}\tilde{x}_p &= \tilde{b}_1 \\ \tilde{a}_{21}\tilde{x}_1 + \tilde{a}_{22}\tilde{x}_2 + \cdots + \tilde{a}_{2p}\tilde{x}_p &= \tilde{b}_2 \\ &\vdots \\ \tilde{a}_{p1}\tilde{x}_1 + \tilde{a}_{p2}\tilde{x}_2 + \cdots + \tilde{a}_{pp}\tilde{x}_p &= \tilde{b}_p. \end{aligned} \tag{5.30}$$

In matrix form, the above FSLE (5.30) may be written as

$$\begin{pmatrix} \tilde{a}_{11} & \tilde{a}_{12} & \cdots & \tilde{a}_{1p} \\ \tilde{a}_{21} & \tilde{a}_{22} & \cdots & \tilde{a}_{2p} \\ \vdots & \vdots & \ddots & \vdots \\ \tilde{a}_{p1} & \tilde{a}_{p2} & \cdots & \tilde{a}_{pp} \end{pmatrix} \cdot \begin{pmatrix} \tilde{x}_1 \\ \tilde{x}_2 \\ \vdots \\ \tilde{x}_p \end{pmatrix} = \begin{pmatrix} \tilde{b}_1 \\ \tilde{b}_2 \\ \vdots \\ \tilde{b}_p \end{pmatrix}. \tag{5.31}$$

Note 5.4

Suppose all the elements of the fuzzy system (5.30) are in the form of triangular fuzzy numbers as

$$(\tilde{a}_{mn}) = (\alpha_{mn}, \beta_{mn}, \gamma_{mn}), \quad (\tilde{x}_n) = (x_n, y_n, z_n), \quad \text{and} \quad [b_m] = (d_m, e_m, f_m).$$

Thus, the fuzzy linear system (5.30) in terms of triangular fuzzy numbers may be presented as

$$\begin{aligned}
&(\alpha_{11}, \beta_{11}, \gamma_{11})(x_1, y_1, z_1) + (\alpha_{12}, \beta_{12}, \gamma_{12})(x_2, y_2, z_2) + \cdots + (\alpha_{1p}, \beta_{1p}, \gamma_{1p})(x_p, y_p, z_p) \\
&\qquad = (d_1, e_1, f_1) \\
&(\alpha_{21}, \beta_{21}, \gamma_{21})(x_1, y_1, z_1) + (\alpha_{22}, \beta_{22}, \gamma_{22})(x_2, y_2, z_2) + \cdots + (\alpha_{2p}, \beta_{2p}, \gamma_{2p})(x_p, y_p, z_p) \\
&\qquad = (d_2, e_2, f_2) \\
&\qquad \vdots \\
&(\alpha_{p1}, \beta_{p1}, \gamma_{p1})(x_1, y_1, z_1) + (\alpha_{p2}, \beta_{p2}, \gamma_{p2})(x_2, y_2, z_2) + \cdots + (\alpha_{pp}, \beta_{pp}, \gamma_{pp})(x_p, y_p, z_p) \\
&\qquad = (d_p, e_p, f_p)
\end{aligned} \tag{5.32}$$

In a similar fashion, FSLE in terms of trapezoidal and Gaussian fuzzy numbers can also be defined.

5.5 SOLUTION OF FSLE

Let us consider a $p \times p$ FSLE in its matrix form as

$$\begin{pmatrix} \tilde{a}_{11} & \tilde{a}_{12} & \cdots & \tilde{a}_{1p} \\ \tilde{a}_{21} & \tilde{a}_{22} & \cdots & \tilde{a}_{2p} \\ \vdots & \vdots & \ddots & \vdots \\ \tilde{a}_{p1} & \tilde{a}_{p2} & \cdots & \tilde{a}_{pp} \end{pmatrix} \cdot \begin{pmatrix} \tilde{x}_1 \\ \tilde{x}_2 \\ \vdots \\ \tilde{x}_p \end{pmatrix} = \begin{pmatrix} \tilde{b}_1 \\ \tilde{b}_2 \\ \vdots \\ \tilde{b}_p \end{pmatrix}. \tag{5.33}$$

Comparing the above system (5.33) with the general form of FSLE $\tilde{A}\tilde{x} = \tilde{b}$, we may have the fuzzy coefficient matrix $\tilde{A}$ and right-hand side fuzzy column vector as

$$\tilde{A} = \begin{pmatrix} \tilde{a}_{11} & \tilde{a}_{12} & \cdots & \tilde{a}_{1p} \\ \tilde{a}_{21} & \tilde{a}_{22} & \cdots & \tilde{a}_{2p} \\ \vdots & \vdots & \ddots & \vdots \\ \tilde{a}_{p1} & \tilde{a}_{p2} & \cdots & \tilde{a}_{pp} \end{pmatrix} \quad \text{and} \quad \tilde{b} = \begin{pmatrix} \tilde{b}_1 \\ \tilde{b}_2 \\ \vdots \\ \tilde{b}_p \end{pmatrix}. \tag{5.34}$$

Here, all the elements of $\tilde{A}$ and $\tilde{b}$ (that is, $\tilde{a}_{mn}$ and $\tilde{b}_m$ for $n, m = 1, 2, \ldots, p$) are in the form of fuzzy numbers viz. triangular, trapezoidal, or Gaussian fuzzy numbers.

Now, the above fuzzy matrix and vector are parameterized by adopting $a-$cut approach (as given in Section 4.1 of Chapter 4) and transformed into the standard interval (depending upon which type of fuzzy number) with parameter $a \in [0, 1]$. Thus, the fuzzy parametric coefficient matrix $\tilde{A}(a)$ and fuzzy parametric column vector $\tilde{b}(a)$ may be written as

$$\tilde{A}(a) = \begin{pmatrix} \left[\underline{a_{11}(a)}, \overline{a_{11}(a)}\right] & \left[\underline{a_{12}(a)}, \overline{a_{12}(a)}\right] & \cdots & \left[\underline{a_{1p}(a)}, \overline{a_{1p}(a)}\right] \\ \left[\underline{a_{21}(a)}, \overline{a_{21}(a)}\right] & \left[\underline{a_{22}(a)}, \overline{a_{22}(a)}\right] & \cdots & \left[\underline{a_{2p}(a)}, \overline{a_{2p}(a)}\right] \\ \vdots & \vdots & \ddots & \vdots \\ \left[\underline{a_{p1}(a)}, \overline{a_{p1}(a)}\right] & \left[\underline{a_{p2}(a)}, \overline{a_{p2}(a)}\right] & \cdots & \left[\underline{a_{pp}(a)}, \overline{a_{pp}(a)}\right] \end{pmatrix},$$

$$\tilde{b}(a) = \begin{pmatrix} \left[\underline{b_1(a)}, \overline{b_1(a)}\right] \\ \left[\underline{b_2(a)}, \overline{b_2(a)}\right] \\ \vdots \\ \left[\underline{b_p(a)}, \overline{b_p(a)}\right] \end{pmatrix}. \tag{5.35}$$

Further, the above fuzzy parametric matrix and column vector are again converted into their respective affine form representations adopting the technique explained in Section 3.3 of Chapter 3. We assume that $\hat{A}(a, \varepsilon_{mn})$ and $\hat{b}(a, \varepsilon_{bm})$ denote the respective fuzzy-affine coefficient matrix and fuzzy-affine column vector, respectively, and then they may be expressed as

$$\hat{A}\,(a, \varepsilon_{mn}) = \begin{pmatrix} \hat{a}_{11}\,(a, \varepsilon_{11}) & \hat{a}_{12}\,(a, \varepsilon_{12}) & \cdots & \hat{a}_{1p}\,(a, \varepsilon_{1p}) \\ \hat{a}_{21}\,(a, \varepsilon_{21}) & \hat{a}_{22}\,(a, \varepsilon_{22}) & \cdots & \hat{a}_{2p}\,(a, \varepsilon_{2p}) \\ \vdots & \vdots & \ddots & \vdots \\ \hat{a}_{p1}\,(a, \varepsilon_{p1}) & \hat{a}_{p2}\,(a, \varepsilon_{p2}) & \cdots & \hat{a}_{pp}\,(a, \varepsilon_{pp}) \end{pmatrix} \quad \text{and}$$

$$\hat{b}\,(a, \varepsilon_{bm}) = \begin{pmatrix} \hat{b}_1\,(a, \varepsilon_{b1}) \\ \hat{b}_2\,(a, \varepsilon_{b2}) \\ \vdots \\ \hat{b}_p\,(a, \varepsilon_{bp}) \end{pmatrix}. \tag{5.36}$$

where $a \in [0, 1]$ is the fuzzy parameter and $\varepsilon_{mn}, \varepsilon_{bm} \in \mathbb{D} = [-1, 1]$ for $m, n = 1, 2, \ldots, p$ are noise symbols (distinct from each other) of each element of the matrix and column vector.

Moreover, the elements are of the form

$$\hat{a}_{mn}(a, \varepsilon_{mn}) = \frac{1}{2}\left(\underline{a_{mn}(a)} + \overline{a_{mn}(a)}\right) + \frac{1}{2}\left(\overline{a_{mn}(a)} - \underline{a_{mn}(a)}\right)\varepsilon_{mn}, \quad \text{for } 1 \leq m, n \leq p; \tag{5.37a}$$

$$\hat{b}_{bm}(a, \varepsilon_{bm}) = \frac{1}{2}\left(\underline{b_{bm}(a)} + \overline{b_{bm}(a)}\right) + \frac{1}{2}\left(\overline{b_{bm}(a)} - \underline{b_{bm}(a)}\right)\varepsilon_{bm}, \quad \text{for } 1 \leq m \leq p. \tag{5.37b}$$

Thus, the $p \times p$ FSLE (5.33) is converted into a fuzzy-affine system of linear equations as follows:

$$\hat{A}(a, \varepsilon_{mn})\,\hat{x}(a, \varepsilon^*) = \hat{b}(a, \varepsilon_{bm}), \tag{5.38}$$

where ε^* may be either a newly generated noise symbol or a function of existing noise symbols ε_{mn} and ε_{bm} for $m, n = 1, 2, \ldots, p$. Moreover, $\hat{x}(a, \varepsilon^*)$ is the corresponding fuzzy-affine solution vector of the fuzzy-affine linear system (5.38).

Now substituting the fuzzy-affine coefficient matrix $\hat{A}(a, \varepsilon_{mn})$ and fuzzy-affine right-hand side column vector $\hat{b}(a, \varepsilon_{bm})$ in the system (5.38), we may have

$$\sum_{n=1}^{p}\left\{\frac{1}{2}\left(\underline{a_{mn}(a)} + \overline{a_{mn}(a)}\right) + \frac{1}{2}\left(\overline{a_{mn}(a)} - \underline{a_{mn}(a)}\right)\varepsilon_{mn}\right\}$$
$$\{\hat{x}_n(a, \varepsilon^*)\} = \frac{1}{2}\left(\underline{b_{bm}(a)} + \overline{b_{bm}(a)}\right) + \frac{1}{2}\left(\overline{b_{bm}(a)} - \underline{b_{bm}(a)}\right)\varepsilon_{bm} \quad \text{for } m = 1, \ldots, p. \tag{5.39}$$

The above system (5.39) is now solved symbolically, having different parameters viz. fuzzy parameter (a) and noise symbols (ε_{mn} and ε_{bm} for $m, n = 1, 2, \ldots, p$) to obtain the fuzzy-affine solution vector $\hat{x}(a, \varepsilon^*)$.

Since every consisting noise symbols vary from -1 to 1, the lower and upper bounds of the fuzzy parametric solution vector $\tilde{x}_n(a) = [\underline{x_n(a)}, \overline{x_n(a)}]$ for $n = 1, 2, \ldots, p$ may be determined as follows:

$$\left.\begin{aligned} \underline{x_n(a)} &= \min_{\varepsilon^* \in [-1,1]} \hat{x}_n(a, \varepsilon^*) \\ \overline{x_n(a)} &= \max_{\varepsilon^* \in [-1,1]} \hat{x}_n(a, \varepsilon^*) \end{aligned}\right\}, \quad \text{for } n = 1, 2, \ldots, p. \tag{5.40}$$

Finally, all the fuzzy solutions of the FSLE given in Eq. (5.33) can be evaluated by varying the fuzzy parameter "a" from 0 to 1.

Example 5.5 Use fuzzy-affine arithmetic to compute the fuzzy solutions of the following FSLE (parameters are in terms of trapezoidal fuzzy numbers) and plot the corresponding trapezoidal fuzzy solutions.

$$(9, 9.5, 10.5, 11)\tilde{x}_1 + (-4.1, -4.05, -3.95, -3.9)\tilde{x}_2 = (26, 27, 29, 30)$$
$$(-4.1, -4.05, -3.95, -3.9)\tilde{x}_1 + (15, 15.25, 15.75, 16)\tilde{x}_2 = (35, 36.5, 37.5, 39).$$

Solution: Comparing the given system of equations with the general form of FSLE $\tilde{A}\tilde{x} = \tilde{b}$, the fuzzy coefficient matrix and fuzzy right-hand side column vector may be given as

$$\tilde{A} = \begin{bmatrix} (9, 9.5, 10.5, 11) & (-4.1, -4.05, -3.95, -3.9) \\ (-4.1, -4.05, -3.95, -3.9) & (15, 15.25, 15.75, 16) \end{bmatrix} \quad \text{and}$$

$$\tilde{b} = \begin{bmatrix} (26, 27, 29, 30) \\ (35, 36.5, 37.5, 39) \end{bmatrix}.$$

Utilizing $a-$cut technique, the above trapezoidal fuzzy matrices are parameterized (using the procedure given in Section 4.1 of Chapter 4) as follows:

$$\tilde{A}(a) = \begin{pmatrix} [9 + 0.5a, 11 - 0.5a] & [-4.1 + 0.05a, -3.9 - 0.05a] \\ [-4.1 + 0.05a, -3.9 - 0.05a] & [15 + 0.25a, 16 - 0.25a] \end{pmatrix} \quad \text{and}$$

$$\tilde{b}(a) = \begin{pmatrix} [26 + a, 30 - a] \\ [35 + 1.5a, 39 - 1.5a] \end{pmatrix}, \quad \text{where} \quad a \in [0, 1].$$

Further, the fuzzy parametric matrices are transformed into fuzzy-affine matrices by converting each of the elements of the matrices into their affine form representations as

$$\hat{A}\,(a, \varepsilon_i) = \begin{pmatrix} 10 + (1 - 0.5a)\varepsilon_1 & -4 + (0.1 - 0.05a)\varepsilon_2 \\ -4 + (0.1 - 0.05a)\varepsilon_3 & 15.5 + (0.5 - 0.25a)\varepsilon_4 \end{pmatrix} \quad \text{for} \quad i = 1, 2, 3, 4, \quad \text{and}$$

$$\hat{b}\,(a, \varepsilon_i) = \begin{pmatrix} 28(2 - a)\varepsilon_5 \\ 37 + (2 - 1.5a)\varepsilon_6 \end{pmatrix} \quad \text{for} \quad i = 5, 6,$$

where $\varepsilon_i \in \mathbb{D} = [-1, 1]$ for $i = 1, 2, \ldots, 6$ are noise symbols distinct from each other.

Thus, the 2×2 FSLE is transformed into a fuzzy-affine system $\hat{A}(a, \varepsilon_i)\hat{x}(a, \varepsilon^*) = \hat{b}(a, \varepsilon_i)$. Here $\hat{x}(a, \varepsilon^*)$ is the fuzzy-affine solution vector and ε^* is the function of existing noise symbols ε_i for $i = 1, 2, \ldots, 6$. The solution vector $\hat{x}(a, \varepsilon^*)$ is evaluated by solving the fuzzy-affine system symbolically in which the solution contains several parameters such as fuzzy parameter (a) and noise symbols (ε_i for $i = 1, 2, \ldots, 6$). By adopting the proposed method, the fuzzy solutions of the given FSLE are as depicted in Figs. 5.2 and 5.3.

Moreover, for some values viz. $a = 0, 0.2, 0.5, 0.7, 0.9$, and 1, the solution bounds of the fuzzy solution vectors are as listed in Table 5.5.

Example 5.6 Find the fuzzy solution plots of the 3×3 FSLE $\tilde{A}\tilde{x} = \tilde{b}$ given as follows:

$$\begin{bmatrix} (0.1, 0.3, 0.9) & (1, 1.5, 1.9) & (0.11, 0.7, 0.9) \\ (0.1, 0.13, 0.2) & (0.11, 0.15, 0.2) & (6, 6.075, 6.2) \\ (5, 5.15, 5.4) & (0.1, 0.3, 0.4) & (0.11, 0.2, 0.4) \end{bmatrix} \cdot \begin{pmatrix} \tilde{x}_1 \\ \tilde{x}_2 \\ \tilde{x}_3 \end{pmatrix} = \begin{bmatrix} (0.01, 0.08, 0.2) \\ (0.11, 0.17, 0.2) \\ (0.1, 0.15, 0.2) \end{bmatrix}.$$

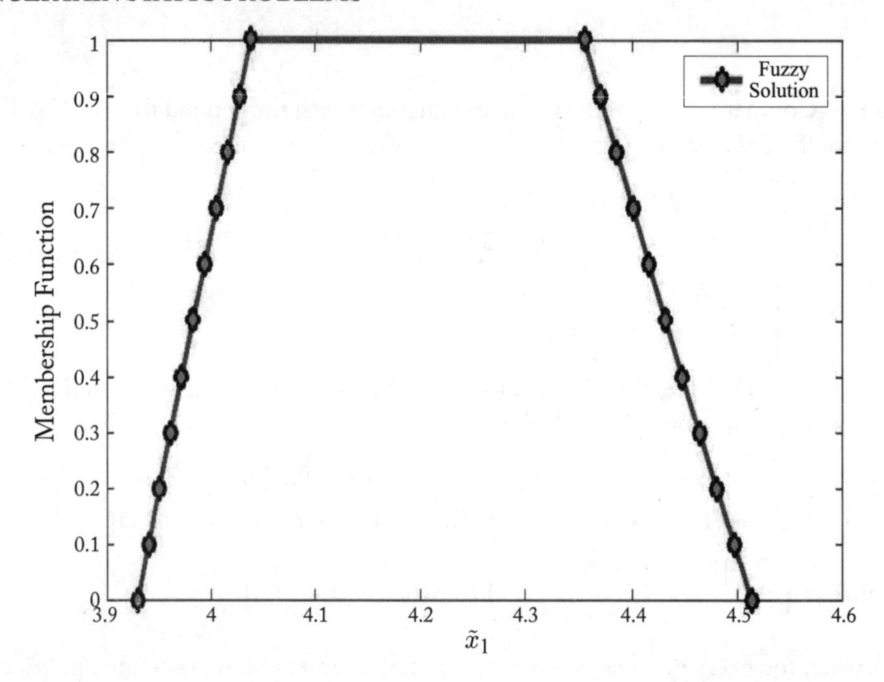

Figure 5.2: Fuzzy solution plot of $\tilde{x}_1$ for Example 5.5.

Table 5.5: Fuzzy solution bounds of Example 5.5 for different values of α

Solution	Lower and Upper Bounds	$\alpha = 0$	$\alpha = 0.2$	$\alpha = 0.5$	$\alpha = 0.7$	$\alpha = 0.9$	$\alpha = 1$ $(\underline{x_n} = \overline{x_n})$
$\tilde{x}_1$	$\underline{x_1}$	3.9312	3.9518	3.9835	4.0052	4.0274	4.0386
	$\overline{x_1}$	4.5139	4.4806	4.4322	4.4010	4.3705	4.3556
$\tilde{x}_2$	$\underline{x_2}$	3.3938	3.3938	3.3940	3.3944	3.3951	3.3957
	$\overline{x_2}$	3.5671	3.5632	3.5578	3.5545	3.5516	3.5502

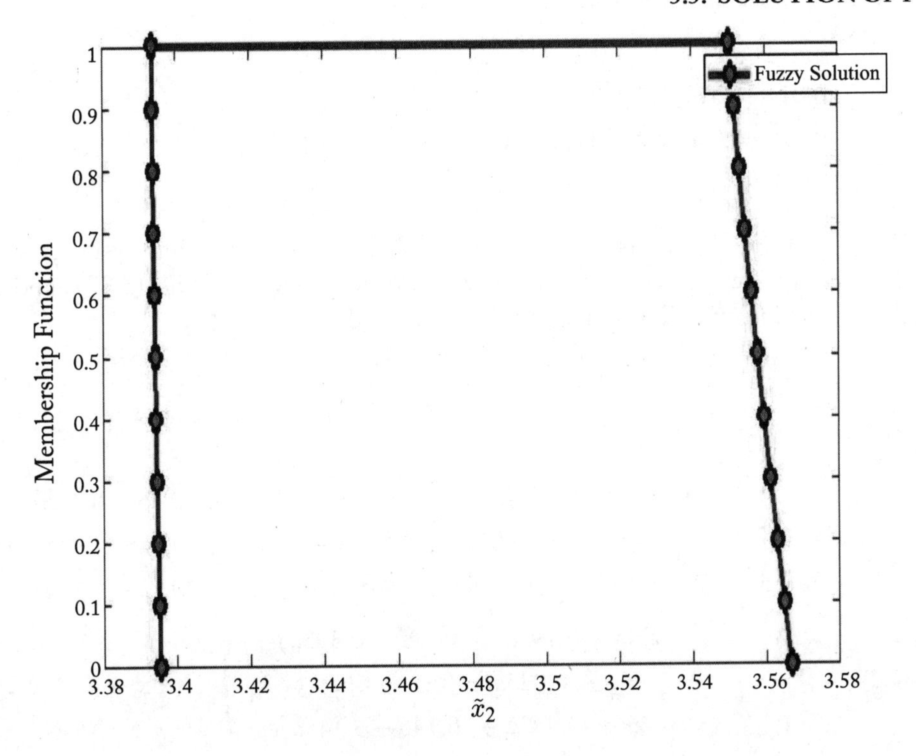

Figure 5.3: Fuzzy solution plot of $\tilde{x}_2$ for Example 5.5.

Solution: Here $\tilde{A}\tilde{x} = \tilde{b}$ is the FSLE, where

$$\tilde{A} = \begin{bmatrix} (0.1, 0.3, 0.9) & (1, 1.5, 1.9) & (0.11, 0.7, 0.9) \\ (0.1, 0.13, 0.2) & (0.11, 0.15, 0.2) & (6, 6.075, 6.2) \\ (5, 5.15, 5.4) & (0.1, 0.3, 0.4) & (0.11, 0.2, 0.4) \end{bmatrix} \text{ and } \tilde{b} = \begin{bmatrix} (0.01, 0.08, 0.2) \\ (0.11, 0.17, 0.2) \\ (0.1, 0.15, 0.2) \end{bmatrix}.$$

Since the consisting elements of the above system are in the form of triangular fuzzy numbers, the fuzzy coefficient matrix and fuzzy column vector may be parameterized as follows:

$$\tilde{A}(a) = \begin{pmatrix} [0.1 + 0.2a, 0.9 - 0.6a] & [1 + 0.5a, 1.9 - 0.4a] & [0.11 + 0.59a, 0.9 - 0.2a] \\ [0.1 + 0.03a, 0.2 - 0.07a] & [0.11 + 0.04a, 0.2 - 0.05a] & [6 + 0.075a, 6.2 - 0.125a] \\ [5 + 0.15a, 5.4 - 0.25a] & [0.1 + 0.2a, 0.4 - 0.1a] & [0.11 + 0.09a, 0.4 - 0.2a] \end{pmatrix}$$

and

$$\tilde{b}(a) = \begin{pmatrix} [0.01 + 0.07a, 0.2 - 0.12a] \\ [0.11 + 0.06a, 0.2 - 0.03a] \\ [0.1 + 0.05a, 0.2 - 0.05a] \end{pmatrix}, \quad \text{where} \quad a \in [0, 1].$$

Further, in order to transform the 3×3 FSLE into a fuzzy-affine system, we have to convert the matrix elements into the fuzzy-affine forms as given in Eqs. (5.37a) and (5.37b). Then, the required fuzzy-affine system may be expressed as

$$\hat{A}(a, \varepsilon_i)\, \hat{x}(a, \varepsilon^*) = \hat{b}(a, \varepsilon_i),$$

where

$$\hat{A}(a, \varepsilon_i) = \begin{pmatrix} 0.5 - 0.2a + (0.4 - 0.4a)\varepsilon_1 & 1.45 + 0.05a + (0.45 - 0.45a)\varepsilon_2 \\ 0.15 - 0.02a + (0.05 - 0.05a)\varepsilon_4 & 0.155 - 0.005a + (0.045 - 0.045a)\varepsilon_5 \\ 5.2 - 0.05a + (0.2 - 0.2a)\varepsilon_7 & 0.25 + 0.05a + (0.15 - 0.15a)\varepsilon_8 \end{pmatrix.$$

$$\begin{matrix} 0.505 + 0.195a + (0.395 - 0.395a)\varepsilon_3 \\ 6.1 - 0.025a + (0.1 - 0.1a)\varepsilon_6 \\ 0.255 - 0.055a + (0.145 - 0.145a)\varepsilon_9 \end{matrix} \Bigg), \quad \text{for} \quad i = 1, 2, \ldots, 9.$$

and

$$\hat{b}(a, \varepsilon_i) = \begin{pmatrix} 0.105 - 0.025a + (0.095 - 0.095a)\varepsilon_{10} \\ 0.155 + 0.015a + (0.045 - 0.045a)\varepsilon_{11} \\ 0.15 + (0.05 - 0.05a)\varepsilon_{12} \end{pmatrix}, \quad \text{for} \quad i = 10, 11, 12.$$

Now, utilizing the proposed procedure, the fuzzy-affine solution vector $\hat{x}(a, \varepsilon^*)$ is evaluated symbolically, having different parameters such as fuzzy parameter (a) and noise symbols (ε_i for $i = 1, 2, \ldots, 12$). Finally, as given in Example 5.5, all the fuzzy plots of the fuzzy solution vector are as depicted in Figs. 5.4–5.6.

5.6 EXERCISES

5.1. Find the affine solution vector of the interval system of linear equations $[A][x] = [b]$, where $[A] = \begin{pmatrix} [4, 6] & [5, 8] \\ [6, 7] & [4, 5] \end{pmatrix}$ and $[b] = \begin{pmatrix} [40, 67] \\ [43, 55] \end{pmatrix}$. Further, find the interval bounds of the affine solution vector.

5.2. Evaluate the interval displacement vector ($[s]$ mm) of a static structural system where the respective interval stiffness matrix ($[K]$ N/mm) and interval force vector ($[F]$ N) are

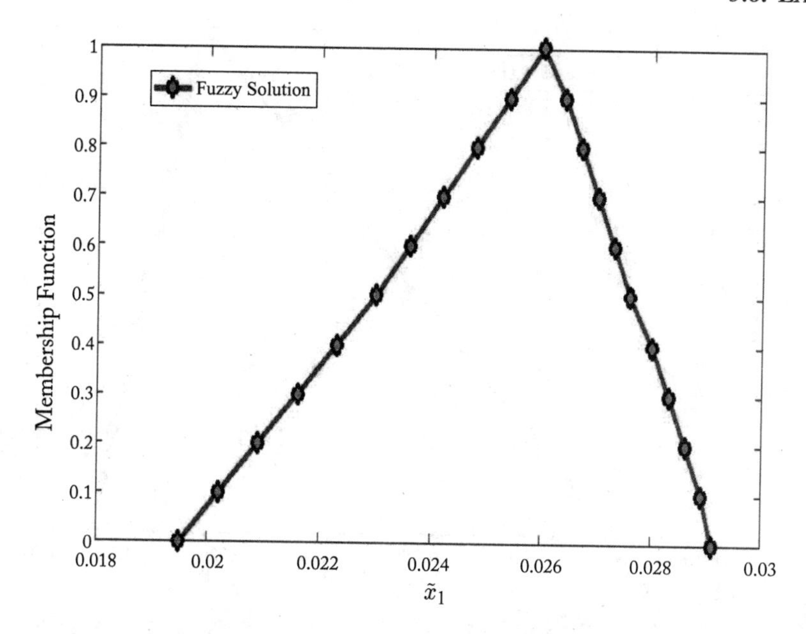

Figure 5.4: Fuzzy solution plot of $\tilde{x}_1$ for Example 5.6.

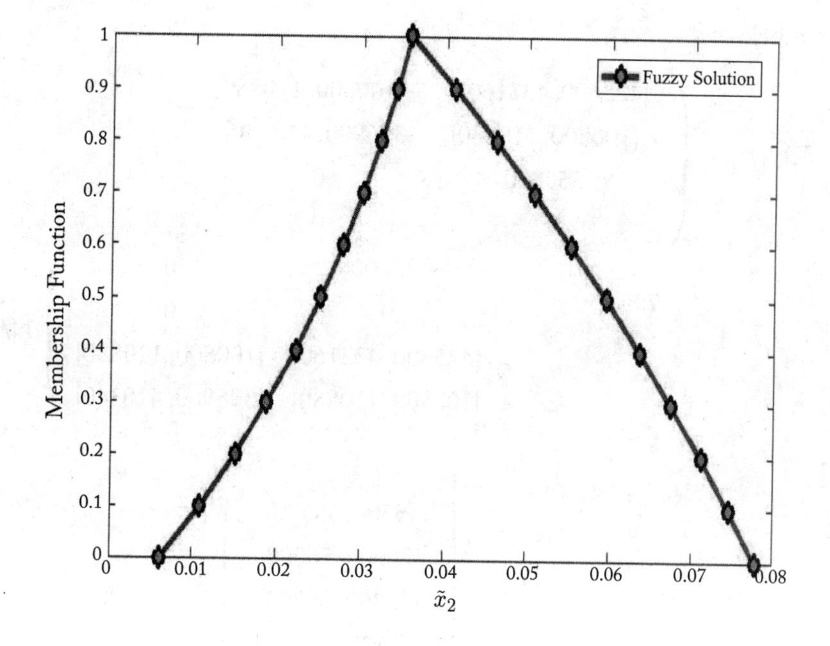

Figure 5.5: Fuzzy solution plot of $\tilde{x}_2$ Example 5.6.

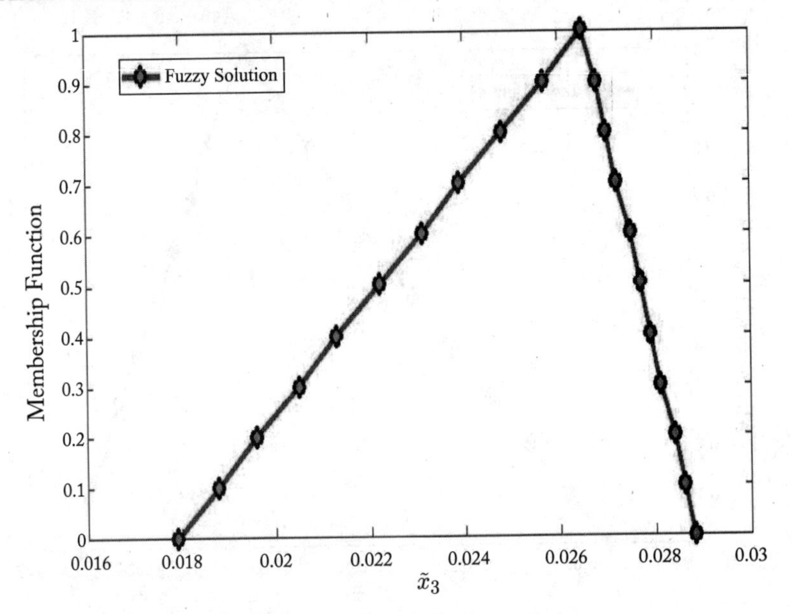

Figure 5.6: Fuzzy solution plot of $\tilde{x}_3$ Example 5.6.

given below:

$$[K] = \begin{pmatrix} [425600, 433160] & -[100800, 110880] & -350000 & 0 \\ -[100800, 110880] & [396900, 410340] & 0 & 0 \\ -350000 & 0 & [425600, 433160] & [100800, 110880] \\ 0 & 0 & [100800, 110880] & [396900, 410340] \end{pmatrix} \text{N/mm}$$

and

$$[F] = \begin{Bmatrix} [5500, 35500] \\ [26000, 56000] \\ [36250, 66250] \\ -[15750, 45750] \end{Bmatrix} \text{N.}$$

Here the interval linear system of equations is $[K][s] = [F]$.

5.3. Find the trapezoidal fuzzy solution plots for the following 2×2 fully fuzzy system of linear equations:

$$(-2,-1,3,4)\tilde{x}_1 + (-2,0,2,3)\tilde{x}_2 = (-13,2,4,14)$$
$$(1,1.5,2,2)\tilde{x}_1 + (4,4,3,5)\tilde{x}_2 = (-14,-8,-4,0).$$

5.4. Find the static response plots of the three-bar truss given in Example 5.3. Consider the system given in Case I of the example and assume here that the eternal load acting at node 3 is in the form of a triangular fuzzy number, that is, $\tilde{P} = (-160,-150,-140)$. Also, compute the lower and upper bounds of the displacement vector.

5.7 REFERENCES

[1] Akhmerov, R. R., 2005. Interval-affine Gaussian algorithm for constrained systems. *Reliable Computing*, 11(5), 323–341. DOI: 10.1007/s11155-005-0040-5.

[2] Behera, D. and Chakraverty, S., 2015. New approach to solve fully fuzzy system of linear equations using single and double parametric form of fuzzy numbers. *Sadhana*, 40(1):35–49. DOI: 10.1007/s12046-014-0295-9.

[3] Behera, D. and Chakraverty, S., 2012. Solution of fuzzy system of linear equations with polynomial parametric form. *Applications and Applied Mathematics*, 7(2):648–657.

[4] Behera, D. and Chakraverty, S., 2013. Fuzzy finite element analysis of imprecisely defined structures with fuzzy nodal force. *Engineering Applications of Artificial Intelligence*, 26(10):2458–2466. DOI: 10.1016/j.engappai.2013.07.021.

[5] Behera, D. and Chakraverty, S., 2013. Fuzzy finite element based solution of uncertain static problems of structural mechanics. *International Journal of Computer Applications*, 69(15). DOI: 10.5120/11916-8040.

[6] Bhavikatti, S. S., 2005. *Finite Element Analysis*. New Age International Publishers. 82, 83

[7] Chakraverty, S., Hladík, M., and Behera, D., 2017. Formal solution of an interval system of linear equations with an application in static responses of structures with interval forces. *Applied Mathematical Modelling*, 50:105–117. DOI: 10.1016/j.apm.2017.05.010.

[8] Chakraverty, S., Hladík, M., and Mahato, N. R., 2017. A sign function approach to solve algebraically interval system of linear equations for nonnegative solutions. *Fundamenta Informaticae*, 152(1):13–31. DOI: 10.3233/fi-2017-1510.

[9] Chakraverty, S., Sahoo, D. M., and Mahato, N. R., 2019. *Concepts of Soft Computing: Fuzzy and ANN with Programming*. Springer. DOI: 10.1007/978-981-13-7430-2. 79, 80

[10] Dehghan, M., Hashemi, B., and Ghatee, M., 2006. Computational methods for solving fully fuzzy linear systems. *Applied Mathematics and Computation*, 179(1):328–343. DOI: 10.1016/j.amc.2005.11.124.

[11] Dubois, D. J., 1980. *Fuzzy Sets and Systems: Theory and Applications*, 144. Academic Press.

[12] Hladík, M., 2012. Enclosures for the solution set of parametric interval linear systems. *International Journal of Applied Mathematics and Computer Science*, 22(3):561–574. DOI: 10.2478/v10006-012-0043-4.

[13] Jeswal, S. K. and Chakraverty, S., 2019. Connectionist model for solving static structural problems with fuzzy parameters. *Applied Soft Computing*, 78:221–229. DOI: 10.1016/j.asoc.2019.02.025.

[14] Karunakar, P. and Chakraverty, S., 2018. Solving fully interval linear systems of equations using tolerable solution criteria. *Soft Computing*, 22(14):4811–4818. DOI: 10.1007/s00500-017-2668-6.

[15] Skalna, I., 2009. Direct method for solving parametric interval linear systems with non-affine dependencies. In *International Conference on Parallel Processing and Applied Mathematics*, pages 485–494, Springer, Berlin, Heidelberg. DOI: 10.1007/978-3-642-14403-5_51.

[16] Soares, R. D. P., 2013. Finding all real solutions of nonlinear systems of equations with discontinuities by a modified affine arithmetic. *Computers and Chemical Engineering*, 48:48–57. DOI: 10.1016/j.compchemeng.2012.08.002.

CHAPTER 6

Uncertain Linear Dynamic Problems

The dynamic analysis of various science and engineering problems with different material and geometric properties lead to linear eigenvalue problems (LEPs) such as the generalized eigenvalue problem (GEP) and standard eigenvalue problem (SEP). In general, the material and geometric properties are assumed to be in the form of crisp (or exact). However, due to several errors and insufficient or incomplete information of data, uncertainties are assumed to be present in the material and geometric properties. Traditionally, these uncertainties are modeled through probabilistic approaches, which are unable to deliver efficient and reliable solutions without a sufficient amount of experimental data. Thus, these uncertain material and geometric properties may be modeled through closed intervals or convex normalized fuzzy sets. In this regard, efficient handling of these eigenvalue problems in an uncertain environment is a challenging and important task to deal with.

Linear eigenvalue problems (GEP or SEP) have a wide variety of applications in several scientific and engineering fields viz. structural mechanics, control theory, fluid dynamics and electrical circuitry. For instance, the linear eigenvalue problem plays a major role in structural dynamic problems.

For this, let us consider an l degrees-of-freedom spring-mass structural system as shown in Fig. 6.1. Here, m_i (kg) for $i = 1, \ldots, l$ are the masses and k_i (N/m) for $i = 1, \ldots, l + 1$ are the stiffnesses of the springs.

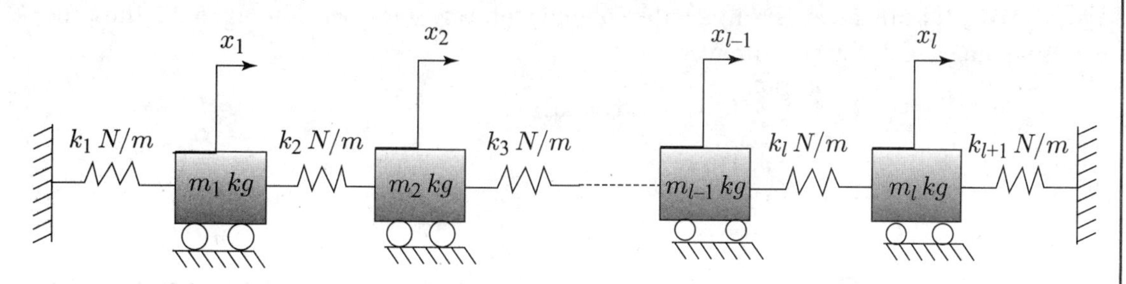

Figure 6.1: Spring-mass structural system of l degrees-of-freedom.

Thus, the equation of motion for the l degrees-of-freedom spring-mass structural system subject to ambient vibration may be obtained as [Chakraverty and Behera (2017) [2]]

$$Kv = -M\ddot{v}, \tag{6.1}$$

where K and M are the respective global stiffness and mass matrices given as follow:

$$K = \begin{pmatrix} k_1 + k_2 & -k_2 & 0 & \cdots & 0 \\ -k_2 & k_2 + k_3 & -k_3 & \vdots & 0 \\ \vdots & \ddots & \ddots & \ddots & \vdots \\ 0 & \vdots & -k_{l-1} & k_{l-1} + k_l & -k_l \\ 0 & \cdots & 0 & -k_l & k_l + k_{l+1} \end{pmatrix} \quad \text{and}$$

$$M = \begin{pmatrix} m_1 & 0 & \cdots & 0 & 0 \\ 0 & m_2 & \cdots & 0 & 0 \\ 0 & \vdots & \ddots & \vdots & \vdots \\ 0 & 0 & \cdots & m_{l-1} & 0 \\ 0 & 0 & \cdots & 0 & m_l \end{pmatrix}. \tag{6.2}$$

Further, v and $\ddot{v}$ are the deflection and acceleration vectors, respectively. For simple harmonic motion, by substituting $v = xe^{i\omega t}$ in Eq. (6.1), the governing equation of motion may be reduced to a GEP

$$Kx = \lambda Mx, \tag{6.3}$$

where $\lambda = \omega^2$ stands for the eigenvalue and x is the corresponding eigenvector of the above problem. Here, the coefficient matrices of the GEP are given in Eq. (6.2).

Moreover, the dynamic analysis of column stiffness of multi-story frame structure also leads to the generalized eigenvalue problem $Kx = \lambda Mx$ (as given in Eq. (6.3)).

For simplicity, let us consider a two-story frame structure [Chakraverty and Behera (2014) [1]] as shown in Fig. 6.2. Here, m_1 (kg) and m_2 (kg) are the floor masses and k_1 (N/m), k_2 (N/m), k_3 (N/m) and k_4 (N/m) are the column stiffnesses (as labeled in Fig. 6.2). Thus, the corresponding GEP may be written as

$$Kx = \lambda Mx,$$

where

$$K = \begin{pmatrix} k_1 + k_2 + k_3 + k_4 & -(k_3 + k_4) \\ -(k_3 + k_4) & (k_3 + k_4) \end{pmatrix} \quad \text{and} \quad M = \begin{pmatrix} m_1 & 0 \\ 0 & m_2 \end{pmatrix} \tag{6.4}$$

are 2×2 mass and stiffness matrices, respectively. Further, $\{x\}$ is a 2×1 displacement vector, and the scalar λ is the natural frequency.

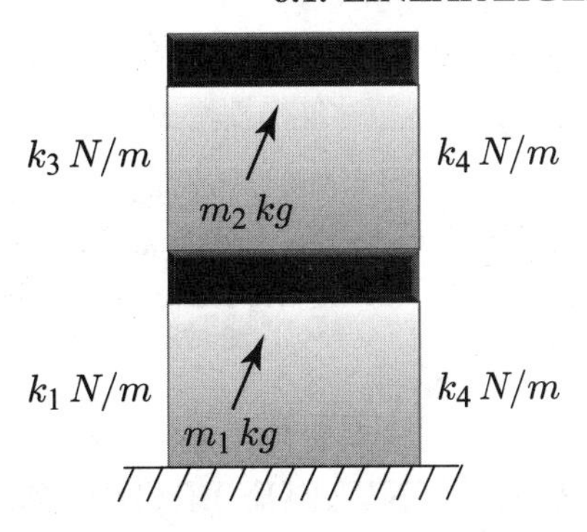

Figure 6.2: Two-story frame structural system.

Therefore, the generalized (or standard) eigenvalue problem plays a vital role in various application problems. Keeping this in view, evaluating the solutions of generalized (or standard) eigenvalue problems become a very important and challenging task for many researchers.

6.1 LINEAR EIGENVALUE PROBLEM (LEP)

Suppose $L(\lambda)$ is a linear matrix-valued function such that $L : \Delta \to C$, where $\Delta \subseteq C$ is an open set. Then, the LEP is the finding of the ordered pairs $(x, \lambda) \in C \times \Delta$ (for $x \neq 0$) such that

$$L(\lambda)x = 0. \tag{6.5}$$

For such ordered pair (x, λ), λ is the known as the eigenvalue and x is the corresponding eigenvector of λ.

The linear eigenvalue problem is classified into two parts viz. GEP and SEP. In the following sections, the definitions and procedures to solve the eigenvalue problems with respect to both interval and fuzzy parameters are provided.

6.2 CRISP GENERALIZED EIGENVALUE PROBLEM (CGEP)

When the parameters of the GEP are crisp (or exact) numbers, it may be referred to as a crisp generalized eigenvalue problem (CGEP). Thus, the general form of the CGEP may be defined as

$$L(\lambda)x = (G - \lambda H)x = 0. \tag{6.6}$$

It may be further written as

$$Gx = \lambda Hx, \tag{6.7}$$

where the above two coefficient matrices $G = (g_{ij})$ and $H = (h_{ij})$ for $i, j = 1, 2, \ldots, l$ are crisp square matrices of order $l \times l$.

Note 6.1

The SEP is a special case of GEP when the right-hand coefficient matrix (H) of the GEP is considered to be an identity matrix (I) having the same dimension as the left-hand coefficient matrix (G) of the GEP.

6.2.1 CRISP STANDARD EIGENVALUE PROBLEM (CSEP)

For $H = I$, Eq. (6.6) may be referred to as a crisp standard eigenvalue problem and is defined as

$$L(\lambda)x = (S - \lambda I)x = 0, \tag{6.8}$$

where the coefficient matrix $S = (s_{ij})$ for $i, j = 1, 2, \ldots, l$ is a crisp square matrix of order $l \times l$ and I is a $l \times l$ identity matrix.

Similarly, CSEP may be further written as

$$Sx = \lambda x. \tag{6.9}$$

Note 6.2

A GEP can also be converted into a SEP by multiplying both sides of Eq. (6.7) with H^{-1} (if the inverse exists). That is, $S = H^{-1}G$ for $\det(H) \neq 0$.

6.3 SOLUTION OF CGEP

There exist many well-known methods for the solution of generalized as well as standard eigenvalue problems (in crisp case) viz. determinant search, subspace iteration, vector iteration, Lanczos method and transformation. For the crisp case (that is for CGEP and CSEP), we can directly compute the eigenvalue solutions by using "MATLAB" with the codes "eig(G,H)" and "eig(S)," respectively, in general.

Example 6.3 Let us consider a two-story frame structure as shown in Fig. 6.2. For this example, suppose all the structural parameters are in the form of crisp numbers. Let the floor masses and the column stiffnesses be as follows:

$$m_1 = m_2 = 3600\,(\text{kg}) \quad \text{and} \quad k_1 = k_2 = 5400\,(\text{N/m}), \quad k_3 = k_4 = 3600\,(\text{N/m}).$$

Thus, compute the eigenvalue solutions of the CGEP obtained during the dynamic analysis of the given structural system.

Table 6.1: Crisp eigenvalue solutions (λ) of Example 6.3

i	Eigenvalues (λ_i)
1	6.0
2	1.0

Solution: From Eq. (6.4), the stiffness and mass matrices of the CGEP $Kx = \lambda Mx$ are obtained as

$$K = \begin{pmatrix} 18000 & -7200 \\ -7200 & 7200 \end{pmatrix} \quad \text{and} \quad M = \begin{pmatrix} 3600 & 0 \\ 0 & 3600 \end{pmatrix}.$$

Corresponding coefficient matrices are of the order 2×2, so there exist at most two eigenvalues of the given CGEP. Then, by using any well-known method, all the crisp eigenvalue solutions may be computed. Further, all these solutions (obtained by using MATLAB) are listed in Table 6.1.

Example 6.4 Determine all the eigenvalue solutions of the CGEP $Gx = \lambda Hx$, where the two coefficient matrices are taken as five-dimensional crisp square matrices given as follows:

$$G = \begin{pmatrix} 3835 & -1825 & 0 & 0 & 0 \\ -1825 & 3440 & -1615 & 0 & 0 \\ 0 & -1615 & 3025 & -1410 & 0 \\ 0 & 0 & -1410 & 2615 & -1205 \\ 0 & 0 & 0 & -1205 & 1205 \end{pmatrix}$$

and

$$H = \begin{pmatrix} 30 & 0 & 0 & 0 & 0 \\ 0 & 27 & 0 & 0 & 0 \\ 0 & 0 & 27 & 0 & 0 \\ 0 & 0 & 0 & 25 & 0 \\ 0 & 0 & 0 & 0 & 18 \end{pmatrix}.$$

Solution: Since the given coefficient matrices are of the order 5×5, there are at most five eigenvalues of the given CGEP. In order to compute these eigenvalues, we may solve the given CGEP by using any well-known method; all the solutions (obtained by using MATLAB) are listed in Table 6.2.

Table 6.2: Crisp eigenvalue solutions (λ) of Example 6.4

i	Eigenvalues (λ_i)
1	6.1662
2	44.0780
3	103.5670
4	165.5908
5	219.4201

6.4 INTERVAL GENERALIZED EIGENVALUE PROBLEM (IGEP)

When all the constituting parameters of a GEP are uncertain and may be taken in the form of closed intervals, the GEP is addressed as an interval generalized eigenvalue problem (IGEP). Thus, the IGEP may be defined as to find the interval scalars ($[\lambda]$) and nonzero interval vectors ($[x]$) such that

$$[G][x] = [\lambda][H][x], \tag{6.10}$$

where $[\lambda]$ is known as the interval eigenvalue and $[x]$ is the corresponding interval eigenvector. Here, the above two coefficient matrices $[G]$ and $[H]$ are the interval square matrices of order $l \times l$ defined as

$$[G] = \begin{pmatrix} [g_{11}] & [g_{12}] & \cdots & [g_{1l}] \\ [g_{21}] & [g_{22}] & \cdots & [g_{2l}] \\ \vdots & \vdots & \ddots & \vdots \\ [g_{l1}] & [g_{l2}] & \cdots & [g_{ll}] \end{pmatrix} \quad \text{and} \quad [H] = \begin{pmatrix} [h_{11}] & [h_{12}] & \cdots & [h_{1l}] \\ [h_{21}] & [h_{22}] & \cdots & [h_{2l}] \\ \vdots & \vdots & \ddots & \vdots \\ [h_{l1}] & [h_{l2}] & \cdots & [h_{ll}] \end{pmatrix}. \tag{6.11}$$

The elements of the above coefficient matrices (given in Eq. (6.11)) are taken in the form of closed intervals as

$$[g_{ij}] = \left[\underline{g_{ij}}, \overline{g_{ij}}\right] \quad \text{and} \quad [h_{ij}] = \left[\underline{h_{ij}}, \overline{h_{ij}}\right], \quad \text{for} \quad i, j = 1, \ldots, l. \tag{6.12}$$

6.4.1 INTERVAL STANDARD EIGENVALUE PROBLEM (ISEP)

In a similar fashion as given in Section 6.2.1, the IGEP (6.10) is referred to as an interval standard eigenvalue problem (ISEP), and the ISEP may be defined as

$$[S][x] = [\lambda][x], \tag{6.13}$$

where the coefficient matrix is an interval square matrix of order $l \times l$ given as follows:

$$[S] = \begin{pmatrix} [s_{11}] & [s_{12}] & \cdots & [s_{1l}] \\ [s_{21}] & [s_{22}] & \cdots & [s_{2l}] \\ \vdots & \vdots & \ddots & \vdots \\ [s_{l1}] & [s_{l2}] & \cdots & [s_{ll}] \end{pmatrix}, \tag{6.14}$$

where

$$[s_{ij}] = \left[\underline{s_{ij}}, \overline{s_{ij}}\right], \quad \text{for} \quad i, j = 1, \ldots, l. \tag{6.15}$$

6.5 SOLUTION OF IGEP

In this section, an affine arithmetic based approach proposed to solve the IGEP (or ISEP). In this regard, let us consider an IGEP as follows:

$$[G][x] = [\lambda][H][x], \tag{6.16}$$

where the above two coefficient matrices $[G]$ and $[H]$ are the interval square matrices of order $l \times l$ defined as

$$[G] = \begin{pmatrix} [g_{11}] & [g_{12}] & \cdots & [g_{1l}] \\ [g_{21}] & [g_{22}] & \cdots & [g_{2l}] \\ \vdots & \vdots & \ddots & \vdots \\ [g_{l1}] & [g_{l2}] & \cdots & [g_{ll}] \end{pmatrix} \quad \text{and} \quad [H] = \begin{pmatrix} [h_{11}] & [h_{12}] & \cdots & [h_{1l}] \\ [h_{21}] & [h_{22}] & \cdots & [h_{2l}] \\ \vdots & \vdots & \ddots & \vdots \\ [h_{l1}] & [h_{l2}] & \cdots & [h_{ll}] \end{pmatrix}. \tag{6.17}$$

Here, the elements of the matrices (6.17) are of the form $[g_{ij}] = \left[\underline{g_{ij}}, \overline{g_{ij}}\right]$ and $[h_{ij}] = \left[\underline{h_{ij}}, \overline{h_{ij}}\right]$, for $i, j = 1, \ldots, l$. Thus, the coefficient matrices may be expressed as

$$[G] = \begin{pmatrix} \left[\underline{g_{11}}, \overline{g_{11}}\right] & \left[\underline{g_{12}}, \overline{g_{12}}\right] & \cdots & \left[\underline{g_{1l}}, \overline{g_{1l}}\right] \\ \left[\underline{g_{21}}, \overline{g_{21}}\right] & \left[\underline{g_{22}}, \overline{g_{22}}\right] & \cdots & \left[\underline{g_{2l}}, \overline{g_{2l}}\right] \\ \vdots & \vdots & \ddots & \vdots \\ \left[\underline{g_{l1}}, \overline{g_{l1}}\right] & \left[\underline{g_{l2}}, \overline{g_{l2}}\right] & \cdots & \left[\underline{g_{ll}}, \overline{g_{ll}}\right] \end{pmatrix}$$

and

$$[H] = \begin{pmatrix} \left[\underline{h_{11}}, \overline{h_{11}}\right] & \left[\underline{h_{12}}, \overline{h_{12}}\right] & \cdots & \left[\underline{h_{1l}}, \overline{h_{1l}}\right] \\ \left[\underline{h_{21}}, \overline{h_{21}}\right] & \left[\underline{h_{22}}, \overline{h_{22}}\right] & \cdots & \left[\underline{h_{2l}}, \overline{h_{2l}}\right] \\ \vdots & \vdots & \ddots & \vdots \\ \left[\underline{h_{l1}}, \overline{h_{l1}}\right] & \left[\underline{h_{l2}}, \overline{h_{l2}}\right] & \cdots & \left[\underline{h_{ll}}, \overline{h_{ll}}\right] \end{pmatrix}. \tag{6.18}$$

Now, to compute the eigenvalue solutions of the given IGEP (6.16), we have adopted an affine arithmetic approach. First, we have to transform the two interval coefficient matrices (6.18) into their affine form representations by using the technique given in Chapter 3. Then, after the conversion, the IGEP is transformed into an affine generalized eigenvalue problem (AGEP), having different parameters in the form of noise symbols. Therefore, the AGEP may be solved as described below.

Suppose, $(G)_c$ and $(G)_\Delta$ are the respective center and half-width (or radius) of the coefficient matrix $[G]$. Similarly, $(H)_c$ and $(H)_\Delta$ are the respective center and half-width of the coefficient matrix $[H]$. Thus, we may have

$$(G)_c = \begin{pmatrix} {g_{11}}^{(0)} & {g_{12}}^{(0)} & \cdots & {g_{1l}}^{(0)} \\ {g_{21}}^{(0)} & {g_{22}}^{(0)} & \cdots & {g_{2l}}^{(0)} \\ \vdots & \vdots & \ddots & \vdots \\ {g_{l1}}^{(0)} & {g_{l2}}^{(0)} & \cdots & {g_{ll}}^{(0)} \end{pmatrix} \text{ and } (G)_\Delta = \begin{pmatrix} {g_{11}}^{(1)} & {g_{12}}^{(1)} & \cdots & {g_{1l}}^{(1)} \\ {g_{21}}^{(1)} & {g_{22}}^{(1)} & \cdots & {g_{2l}}^{(1)} \\ \vdots & \vdots & \ddots & \vdots \\ {g_{l1}}^{(1)} & {g_{l2}}^{(1)} & \cdots & {g_{ll}}^{(1)} \end{pmatrix}; \tag{6.19a}$$

$$(H)_c = \begin{pmatrix} {h_{11}}^{(0)} & {h_{12}}^{(0)} & \cdots & {h_{1l}}^{(0)} \\ {h_{21}}^{(0)} & {h_{22}}^{(0)} & \cdots & {h_{2l}}^{(0)} \\ \vdots & \vdots & \ddots & \vdots \\ {h_{l1}}^{(0)} & {h_{l2}}^{(0)} & \cdots & {h_{ll}}^{(0)} \end{pmatrix} \text{ and } (H)_\Delta = \begin{pmatrix} {h_{11}}^{(1)} & {h_{12}}^{(1)} & \cdots & {h_{1l}}^{(1)} \\ {h_{21}}^{(1)} & {h_{22}}^{(1)} & \cdots & {h_{2l}}^{(1)} \\ \vdots & \vdots & \ddots & \vdots \\ {h_{l1}}^{(1)} & {h_{l2}}^{(1)} & \cdots & {h_{ll}}^{(1)} \end{pmatrix}. \tag{6.19b}$$

where

$$\begin{aligned} {(g)_{ij}}^{(0)} &= \frac{1}{2}\left(\underline{(g)_{ij}} + \overline{(g)_{ij}}\right) \quad \text{and} \quad {(g)_{ij}}^{(1)} = \frac{1}{2}\left(\overline{(g)_{ij}} - \underline{(g)_{ij}}\right); \\ {(h)_{ij}}^{(0)} &= \frac{1}{2}\left(\underline{(h)_{ij}} + \overline{(h)_{ij}}\right) \quad \text{and} \quad {(h)_{ij}}^{(1)} = \frac{1}{2}\left(\overline{(h)_{ij}} - \underline{(h)_{ij}}\right) \quad \text{for} \quad i, j = 1, \ldots, l. \end{aligned}$$

Hence, from Chapter 3, the affine form representation of the interval coefficient matrices (6.18) having different noise symbols are found as follows:

$$\hat{G}\left(\varepsilon_{g,ij}\right) = \begin{pmatrix} g_{11}{}^{(0)} + g_{11}{}^{(1)}\varepsilon_{g,11} & g_{12}{}^{(0)} + g_{12}{}^{(1)}\varepsilon_{g,11} & \cdots & g_{1l}{}^{(0)} + g_{1l}{}^{(1)}\varepsilon_{g,1l} \\ g_{21}{}^{(0)} + g_{21}{}^{(1)}\varepsilon_{g,21} & g_{22}{}^{(0)} + g_{22}{}^{(1)}\varepsilon_{g,22} & \cdots & g_{2l}{}^{(0)} + g_{2l}{}^{(1)}\varepsilon_{g,2l} \\ \vdots & \vdots & \ddots & \vdots \\ g_{l1}{}^{(0)} + g_{l1}{}^{(1)}\varepsilon_{g,l1} & g_{l2}{}^{(0)} + g_{l2}{}^{(1)}\varepsilon_{g,l2} & \cdots & g_{ll}{}^{(0)} + g_{ll}{}^{(1)}\varepsilon_{g,ll} \end{pmatrix}. \tag{6.20a}$$

$$\hat{H}\left(\varepsilon_{h,ij}\right) = \begin{pmatrix} h_{11}{}^{(0)} + h_{11}{}^{(1)}\varepsilon_{h,11} & h_{12}{}^{(0)} + h_{12}{}^{(1)}\varepsilon_{h,11} & \cdots & h_{1l}{}^{(0)} + h_{1l}{}^{(1)}\varepsilon_{h,1l} \\ h_{21}{}^{(0)} + h_{21}{}^{(1)}\varepsilon_{h,21} & h_{22}{}^{(0)} + h_{22}{}^{(1)}\varepsilon_{h,22} & \cdots & h_{2l}{}^{(0)} + h_{2l}{}^{(1)}\varepsilon_{h,2l} \\ \vdots & \vdots & \ddots & \vdots \\ h_{l1}{}^{(0)} + h_{l1}{}^{(1)}\varepsilon_{h,l1} & h_{l2}{}^{(0)} + h_{l2}{}^{(1)}\varepsilon_{h,l2} & \cdots & h_{ll}{}^{(0)} + h_{ll}{}^{(1)}\varepsilon_{h,ll} \end{pmatrix}, \tag{6.20b}$$

where $\varepsilon_{g,ij} \in [-1, 1]$ and $\varepsilon_{h,ij} \in [-1, 1]$ for $i, j = 1, \ldots, l$ are the noise symbols of each element of the above two coefficient matrices (6.20a) and (6.20b), respectively.

Then, the IGEP is converted into its affine form representation and may be called an affine generalized eigenvalue problem (AGEP). It may be denoted as

$$\hat{G}\left(\varepsilon_{g,ij}\right)\hat{x}\left(\varepsilon^*\right) = \hat{\lambda}\left(\varepsilon^*\right)\hat{H}\left(\varepsilon_{h,ij}\right)\hat{x}\left(\varepsilon^*\right), \tag{6.21}$$

where ε^* may be either a newly generated noise symbol or a function of existing noise symbols $\varepsilon_{g,ij}$ and $\varepsilon_{h,ij}$ for $i, j = 1, \ldots, l$. Moreover, $\hat{\lambda}(\varepsilon^*)$ and $\hat{x}(\varepsilon^*)$ are the corresponding affine forms of the interval eigenvalue $[\lambda]$ and interval eigenvector $[x]$ of the IGEP (6.16), respectively.

Here, all elements of the AGEP are in the form of affine representations having different parameters in the form of noise symbols. Hence, the affine eigenvalue solutions in terms of noise symbols are obtained by solving

$$\det\left(\hat{G}\left(\varepsilon_{g,ij}\right) - \hat{\lambda}\hat{H}\left(\varepsilon_{h,ij}\right)\right) = 0 \tag{6.22}$$

symbolically. Thus, all the obtained solutions are in symbolic form with symbols $\varepsilon_{g,ij}$ and $\varepsilon_{h,ij}$ for $i, j = 1, \ldots, l$, which are the required affine eigenvalue solutions $\hat{\lambda}_i(\varepsilon^*)$ for $i = 1, 2, \ldots, l$ of the AGEP (6.22).

Finally, the interval bounds of the eigenvalue solutions $[\lambda_i] = \left[\underline{\lambda_i}, \overline{\lambda_i}\right]$ for $i = 1, 2, \ldots, l$ of the given IGEP (6.16) may be computed by varying each noise symbol from -1 to 1. Thus, we may have

- Lower bounds:

$$\underline{\lambda_i} = \min_{\varepsilon^* \in [-1,1]} \hat{\lambda}_i\left(\varepsilon^*\right); \tag{6.23}$$

- Upper bounds:

$$\overline{\lambda_i} = \max_{\varepsilon^* \in [-1,1]} \hat{\lambda}_i\left(\varepsilon^*\right); \tag{6.24}$$

for $i = 1, 2, \ldots, l$.

Note 6.5

The ISEP $[S][x] = [\lambda][x]$ may be solved similarly as mentioned in the above Section 6.5. First, by converting the interval coefficient matrix $[S]$ into its affine form representation ($\hat{S}(\varepsilon_{ij})$ for $i = 1, 2, \ldots, l$), the ISEP is transformed into an affine standard eigenvalue problem (ASEP) $\hat{S}(\varepsilon_{ij})\hat{x}(\varepsilon^*) = \hat{\lambda}(\varepsilon^*)\hat{x}(\varepsilon^*)$ (as given in Eq. (6.21)). Then, solving $\det\left(\hat{S}(\varepsilon_{ij}) - \hat{\lambda} I\right) = 0$, we get the required solution.

Example 6.6 Determine the lower and upper eigenvalue bounds of the IGEP $[G][x] = [\lambda][H][x]$ (Mahato and Chakraverty (2016b) [9]), where all the coefficient matrices of the problem may be taken as 5×5 interval matrices as follows:

$$[G] = \begin{pmatrix} [3800, 3870] & [-1850, -1800] & 0 & 0 & 0 \\ [-1850, -1800] & [3400, 3480] & [-1630, -1600] & 0 & 0 \\ 0 & [-1630, -1600] & [3000, 3050] & [-1420, -1400] & 0 \\ 0 & 0 & [-1420, -1400] & [2600, 2630] & [-1210, -1200] \\ 0 & 0 & 0 & [-1210, -1200] & [1200, 1210] \end{pmatrix}$$

and

$$[H] = \begin{pmatrix} 30 & 0 & 0 & 0 & 0 \\ 0 & 27 & 0 & 0 & 0 \\ 0 & 0 & 27 & 0 & 0 \\ 0 & 0 & 0 & 25 & 0 \\ 0 & 0 & 0 & 0 & 18 \end{pmatrix}.$$

Table 6.3: Comparison of eigenvalue bounds ($[\lambda]$) of the IGEP for Example 6.6

i	Proposed Method			Mahato and Chakraverty (2016b)			Crisp
	Center (λ_c)	Lower ($\underline{\lambda_i}$)	Upper ($\overline{\lambda_i}$)	Center (λ_c)	Lower ($\underline{\lambda_i}$)	Upper ($\overline{\lambda_i}$)	Eigenvalues (λ)
1	219.42008	219.27481	219.56640	219.42008	217.18678	221.65595	219.4201
2	165.59085	165.21492	165.96822	165.59085	164.00899	167.17109	165.5908
3	103.56705	102.58894	104.55350	103.56705	101.99025	105.15030	103.5670
4	44.07802	42.47619	45.69673	44.07802	42.16398	45.99666	44.0780
5	6.16623	4.81551	7.48922	6.16623	4.67184	7.64548	6.1662

Solution: In order to solve the given IGEP, first of all, we have to transform the two given interval coefficient matrices into their affine form representations. Here, the coefficient matrix $[G]$ is an interval matrix and the second coefficient matrix $[H]$ is crisp. Thus, we have to convert only the matrix $[G]$ into its affine form representation $\hat{G}$ (as given in Eqs. (6.20a) and (6.20b) of Section 6.5) as follows:

$$\hat{G} = \begin{pmatrix} 3835 + 35\varepsilon_1 & -1825 + 25\varepsilon_2 & 0 & 0 & 0 \\ -1825 + 25\varepsilon_3 & 3440 + 40\varepsilon_4 & -1615 + 15\varepsilon_5 & 0 & 0 \\ 0 & -1615 + 15\varepsilon_6 & 3025 + 25\varepsilon_7 & -1410 + 10\varepsilon_8 & 0 \\ 0 & 0 & -1410 + 10\varepsilon_9 & 2615 + 15\varepsilon_{10} & -1205 + 5\varepsilon_{11} \\ 0 & 0 & 0 & -1205 + 5\varepsilon_{12} & 1205 + 5\varepsilon_{13} \end{pmatrix}$$

and

$$\hat{H} = \begin{pmatrix} 30 & 0 & 0 & 0 & 0 \\ 0 & 27 & 0 & 0 & 0 \\ 0 & 0 & 27 & 0 & 0 \\ 0 & 0 & 0 & 25 & 0 \\ 0 & 0 & 0 & 0 & 18 \end{pmatrix},$$

where $\varepsilon_i \in [-1, 1]$ for $i = 1, \ldots, 13$. Therefore, all the interval eigenvalue solutions of the given IGEP are evaluated by adopting the proposed method given in Section 6.5; the interval bounds of the eigenvalue solutions are listed in Table 6.3. Further, comparisons with the solutions of Mahato and Chakraverty (2016b) [9] are also incorporated in this table.

From Table 6.3, it may be observed that the interval eigenvalues of the given IGEP from the proposed method yields tighter bounds compared with the results of Chakraverty and Mahato (2016b) [9]. Further, it may be noted that the crisp eigenvalue solutions obtained in Example 6.6 lie within the lower and upper bounds of the interval solution and coincide with the central eigenvalue solutions.

Example 6.7 As given in Example 6.3, let us now consider the same two-story frame structure (as shown in Fig. 6.2) in an uncertain environment. In this case, all the structural parameters are assumed to be closed intervals. Suppose the interval floor masses and the interval column stiffnesses are taken as follows:

$$[m_1] = [m_2] = [3200, 4000]\,(\text{kg}) \qquad \text{and}$$

$$[k_1] = [k_2] = [5250, 5550]\,(\text{N/m}), \quad [k_3] = [k_4] = [3425, 3775].(\text{N/m}).$$

Thus, compute the interval eigenvalue bounds of the IGEP obtained during the dynamic analysis of the given uncertain structural system.

Solution: The dynamic analysis of the given uncertain structural system (when the uncertainty is considered in the form of closed intervals) leads to IGEP $[K][x] = [\lambda][M][x]$, where the interval stiffness and interval mass matrices are

$$[K] = \begin{pmatrix} [k_1]+[k_2]+[k_3]+[k_4] & -([k_3]+[k_4]) \\ -([k_3]+[k_4]) & ([k_3]+[k_4]) \end{pmatrix} \quad \text{and} \quad [M] = \begin{pmatrix} [m_1] & 0 \\ 0 & [m_2] \end{pmatrix}.$$

Here, the interval stiffness and mass parameters are given. Then, the affine form representation of the given stiffness and mass parameters are found as

$$\begin{aligned} \hat{k}_1 &= 5400 + 150\varepsilon_1\,(\text{N/m}), \\ \hat{k}_2 &= 5400 + 150\varepsilon_2\,(\text{N/m}), \\ \hat{k}_3 &= 3600 + 175\varepsilon_3\,(\text{N/m}), \\ \hat{k}_4 &= 3600 + 175\varepsilon_4\,(\text{N/m}), \quad \text{and} \\ \hat{m}_1 &= 3600 + 400\varepsilon_5\,(\text{kg}), \quad \hat{m}_2 = 3600 + 400\varepsilon_6\,(\text{kg}), \end{aligned}$$

where $\varepsilon_i \in [-1, 1]$ for $i = 1, \ldots, 6$ are different noise symbols for different parameters. Thus, the affine stiffness matrix and affine mass matrix are obtained as follows:

$$\hat{K} = \begin{pmatrix} \hat{k}_1+\hat{k}_2+\hat{k}_3+\hat{k}_4 & -\left(\hat{k}_3+\hat{k}_4\right) \\ -\left(\hat{k}_3+\hat{k}_4\right) & \left(\hat{k}_3+\hat{k}_4\right) \end{pmatrix} \quad \text{and} \quad \hat{M} = \begin{pmatrix} \hat{m}_1 & 0 \\ 0 & \hat{m}_2 \end{pmatrix}.$$

Table 6.4: Eigenvalue bounds ($[\lambda]$) of the IGEP for Example 6.7

i	Interval Eigenvalue Bounds			Crisp Eigenvalues (λ)
	Center (λ_c)	Lower ($\underline{\lambda_i}$)	Upper ($\overline{\lambda_i}$)	
1	6.0	5.617608	6.478272	6.0
2	1.0	0.932392	1.084228	1.0

That is,

$$\hat{K} = \begin{pmatrix} 18000 + 150\varepsilon_1 + 150\varepsilon_2 + 175\varepsilon_3 + 175\varepsilon_4 & -7200 - 175\varepsilon_3 - 175\varepsilon_4 \\ -7200 - 175\varepsilon_3 - 175\varepsilon_4 & 7200 + 175\varepsilon_3 + 175\varepsilon_4 \end{pmatrix}$$

and

$$\hat{M} = \begin{pmatrix} 3600 + 400\varepsilon_5 & 0 \\ 0 & 3600 + 400\varepsilon_6 \end{pmatrix}.$$

Now, by utilizing the proposed approach given in Section 6.5, all the interval eigenvalue bounds are evaluated. The interval bounds of the eigenvalue solutions are listed in Table 6.4.

From Table 6.4, it may be observed that the crisp eigenvalue solutions obtained from Example 6.7 coincide with the central eigenvalue solutions of the given IGEP.

6.6 FUZZY GENERALIZED EIGENVALUE PROBLEM (FGEP)

When all the uncertain parameters of a GEP may be considered in the form of fuzzy numbers, the GEP is referred to as fuzzy generalized eigenvalue problem (FGEP) and may be defined as

$$\tilde{G}\tilde{x} = \tilde{\lambda}\tilde{H}\tilde{x}, \tag{6.25}$$

where $\tilde{\lambda}$ is known as the fuzzy eigenvalue, $\tilde{x}$ is the corresponding nonzero fuzzy eigenvector, and the two coefficient matrices $\tilde{G}$ and $\tilde{H}$ are the fuzzy square matrices of order $l \times l$ given as follows:

$$\tilde{G} = \begin{pmatrix} \tilde{g}_{11} & \tilde{g}_{12} & \cdots & \tilde{g}_{1l} \\ \tilde{g}_{21} & \tilde{g}_{22} & \cdots & \tilde{g}_{2l} \\ \vdots & \vdots & \ddots & \vdots \\ \tilde{g}_{l1} & \tilde{g}_{l2} & \cdots & \tilde{g}_{ll} \end{pmatrix} \quad \text{and} \quad \tilde{H} = \begin{pmatrix} \tilde{h}_{11} & \tilde{h}_{12} & \cdots & \tilde{h}_{1l} \\ \tilde{h}_{21} & \tilde{h}_{22} & \cdots & \tilde{h}_{2l} \\ \vdots & \vdots & \ddots & \vdots \\ \tilde{h}_{l1} & \tilde{h}_{l2} & \cdots & \tilde{h}_{ll} \end{pmatrix}, \tag{6.26}$$

where all the elements of the above fuzzy coefficient matrices are

$$\tilde{G} = (\tilde{g}_{ij}) \quad \text{and} \quad \tilde{H} = \left(\tilde{h}_{ij}\right) \quad \text{for} \quad i, j = 1, \ldots, l. \tag{6.27}$$

These elements are taken in the form of fuzzy numbers (particularly, as TFNs or TrFNs).

6.6.1 FUZZY STANDARD EIGENVALUE PROBLEM (FSEP)

The fuzzy standard eigenvalue problem (FSEP) may be defined as

$$\tilde{S}\tilde{x} = \tilde{\lambda}\tilde{x}, \tag{6.28}$$

where the coefficient matrix is an $l \times l$ fuzzy square matrix expressed as follows:

$$\tilde{S} = \begin{pmatrix} \tilde{s}_{11} & \tilde{s}_{12} & \cdots & \tilde{s}_{1l} \\ \tilde{s}_{21} & \tilde{s}_{22} & \cdots & \tilde{s}_{2l} \\ \vdots & \vdots & \ddots & \vdots \\ \tilde{s}_{l1} & \tilde{s}_{l2} & \cdots & \tilde{s}_{ll} \end{pmatrix}, \tag{6.29}$$

where the elements of the above coefficient matrix $\tilde{S} = (\tilde{s}_{ij})$, for $i, j = 1, \ldots, l$ are in the form of fuzzy numbers.

6.7 SOLUTION OF FGEP

This section contains a fuzzy-affine based technique to handle the FGEP (or FSEP) and evaluates the fuzzy eigenvalue solutions.

Let us consider the FGEP (from Eq. 6.25) given as follows:

$$\tilde{G}\tilde{x} = \tilde{\lambda}\tilde{H}\tilde{x}, \tag{6.30}$$

where the two coefficient matrices $\tilde{G} = (\tilde{g}_{ij})$ and $\tilde{H} = \left(\tilde{h}_{ij}\right)$ for $i, j = 1, 2, \ldots, l$ are fuzzy square matrices of order $l \times l$ in the form of fuzzy numbers such that

$$\tilde{G} = \begin{pmatrix} \tilde{g}_{11} & \tilde{g}_{12} & \cdots & \tilde{g}_{1l} \\ \tilde{g}_{21} & \tilde{g}_{22} & \cdots & \tilde{g}_{2l} \\ \vdots & \vdots & \ddots & \vdots \\ \tilde{g}_{l1} & \tilde{g}_{l2} & \cdots & \tilde{g}_{ll} \end{pmatrix} \quad \text{and} \quad \tilde{H} = \begin{pmatrix} \tilde{h}_{11} & \tilde{h}_{12} & \cdots & \tilde{h}_{1l} \\ \tilde{h}_{21} & \tilde{h}_{22} & \cdots & \tilde{h}_{2l} \\ \vdots & \vdots & \ddots & \vdots \\ \tilde{h}_{l1} & \tilde{h}_{l2} & \cdots & \tilde{h}_{ll} \end{pmatrix}. \tag{6.31}$$

Now, by utilizing the a-cut technique described in Chapter 4, all the elements (in the form of fuzzy numbers) of the above fuzzy coefficient matrices are parameterized. Thus, the fuzzy coefficient matrices are converted into their fuzzy parametric forms and may be expressed as

$$\tilde{G}(a)=\begin{pmatrix} \left[\underline{g_{11}(a)},\overline{g_{11}(a)}\right] & \left[\underline{g_{12}(a)},\overline{g_{12}(a)}\right] & \cdots & \left[\underline{g_{1l}(a)},\overline{g_{1l}(a)}\right] \\ \left[\underline{g_{21}(a)},\overline{g_{21}(a)}\right] & \left[\underline{g_{22}(a)},\overline{g_{22}(a)}\right] & \cdots & \left[\underline{g_{2l}(a)},\overline{g_{2l}(a)}\right] \\ \vdots & \vdots & \ddots & \vdots \\ \left[\underline{g_{l1}(a)},\overline{g_{l1}(a)}\right] & \left[\underline{g_{l2}(a)},\overline{g_{l2}(a)}\right] & \cdots & \left[\underline{g_{ll}(a)},\overline{g_{ll}(a)}\right] \end{pmatrix}; \tag{6.32a}$$

$$\tilde{H}(a)=\begin{pmatrix} \left[\underline{h_{11}(a)},\overline{h_{11}(a)}\right] & \left[\underline{h_{12}(a)},\overline{h_{12}(a)}\right] & \cdots & \left[\underline{h_{1l}(a)},\overline{h_{1l}(a)}\right] \\ \left[\underline{h_{21}(a)},\overline{h_{21}(a)}\right] & \left[\underline{h_{22}(a)},\overline{h_{22}(a)}\right] & \cdots & \left[\underline{h_{2l}(a)},\overline{h_{2l}(a)}\right] \\ \vdots & \vdots & \ddots & \vdots \\ \left[\underline{h_{l1}(a)},\overline{h_{l1}(a)}\right] & \left[\underline{h_{l2}(a)},\overline{h_{l2}(a)}\right] & \cdots & \left[\underline{h_{ll}(a)},\overline{h_{ll}(a)}\right] \end{pmatrix}, \tag{6.32b}$$

for $a \in [0, 1]$.

Therefore, the given FGEP (6.30) is transformed into a fuzzy parametric GEP

$$\tilde{G}(a)\tilde{x}(a)=\tilde{\lambda}(a)\tilde{H}(a)\tilde{x}(a), \tag{6.33}$$

Further, the fuzzy parametric GEP is again converted into its affine form representation by using the procedure given in Chapters 3 and 4. For this, we have to first transform the above fuzzy parametric coefficient matrices (6.32a) and (6.32b) in the form of their fuzzy-affine representations as follows:

$$\hat{G}\left(a,\varepsilon_{g,ij}\right)=\begin{pmatrix} \hat{g}_{11}\left(a,\varepsilon_{g,11}\right) & \hat{g}_{12}\left(a,\varepsilon_{g,12}\right) & \cdots & \hat{g}_{1l}\left(a,\varepsilon_{g,1l}\right) \\ \hat{g}_{21}\left(a,\varepsilon_{g,21}\right) & \hat{g}_{22}\left(a,\varepsilon_{g,22}\right) & \cdots & \hat{g}_{2l}\left(a,\varepsilon_{g,2l}\right) \\ \vdots & \vdots & \ddots & \vdots \\ \hat{g}_{l1}\left(a,\varepsilon_{g,l1}\right) & \hat{g}_{l2}\left(a,\varepsilon_{g,l2}\right) & \cdots & \hat{g}_{ll}\left(a,\varepsilon_{g,ll}\right) \end{pmatrix}; \tag{6.34a}$$

$$\hat{H}\left(a,\varepsilon_{h,ij}\right)=\begin{pmatrix} \hat{h}_{11}\left(a,\varepsilon_{h,11}\right) & \hat{h}_{12}\left(a,\varepsilon_{h,12}\right) & \cdots & \hat{h}_{1l}\left(a,\varepsilon_{h,1l}\right) \\ \hat{h}_{21}\left(a,\varepsilon_{h,21}\right) & \hat{h}_{22}\left(a,\varepsilon_{h,22}\right) & \cdots & \hat{h}_{2l}\left(a,\varepsilon_{h,2l}\right) \\ \vdots & \vdots & \ddots & \vdots \\ \hat{h}_{l1}\left(a,\varepsilon_{h,l1}\right) & \hat{h}_{l2}\left(a,\varepsilon_{h,l2}\right) & \cdots & \hat{h}_{ll}\left(a,\varepsilon_{h,ll}\right) \end{pmatrix}, \tag{6.34b}$$

where $a \in [0, 1]$, $\varepsilon_{g,ij} \in [-1, 1]$ and $\varepsilon_{h,ij} \in [-1, 1]$ for $i, j = 1, \ldots, l$. Here, $\varepsilon_{g,ij}$ and $\varepsilon_{h,ij}$ for $i, j = 1, \ldots, l$ are different noise symbols for each element of the matrices. Hence, the fuzzy parametric GEP (6.33) is converted to a fuzzy-affine GEP as

$$\hat{G}\left(a,\varepsilon_{g,ij}\right)\hat{x}\left(a,\varepsilon^{*}\right)=\hat{\lambda}\left(a,\varepsilon^{*}\right)\hat{H}\left(a,\varepsilon_{h,ij}\right)\hat{x}\left(a,\varepsilon^{*}\right), \tag{6.35}$$

where ε^* may be either a newly generated noise symbol or a function of existing noise symbols $\varepsilon_{g,ij}$ and $\varepsilon_{h,ij}$. Moreover, $\hat{\lambda}(a, \varepsilon^*)$ and $\hat{x}(a, \varepsilon^*)$ are the corresponding fuzzy-affine eigenvalue and the fuzzy-affine eigenvector of the fuzzy-affine GEP (6.35), respectively.

Here, the above fuzzy-affine GEP (6.35) involves several parameters such as the fuzzy parameter (a) and the noise symbols ($\varepsilon_{g,ij}$ and $\varepsilon_{h,ij}$ for $i, j = 1, \ldots, l$), and it may be solved to find the required eigenvalue solutions in fuzzy-affine forms $\hat{\lambda}_i(a, \varepsilon^*)$ for $i = 1, 2, \ldots, l$.

Now, because each noise symbol ($\varepsilon_{g,ij}$ and $\varepsilon_{h,ij}$ for $i, j = 1, \ldots, l$) varies from -1 to 1, the fuzzy parametric eigenvalues $\tilde{\lambda}_i(a) = \left[\underline{\lambda_i(a)}, \overline{\lambda_i(a)}\right]$ for $i = 1, 2, \ldots, l$ of the given fuzzy parametric GEP (6.33) may be computed as follows:

- Lower bounds:

$$\underline{\lambda_i(a)} = \min_{\varepsilon^* \in [-1,1]} \hat{\lambda}_i\left(a, \varepsilon^*\right); \tag{6.36}$$

- Upper bounds:

$$\overline{\lambda_i(a)} = \max_{\varepsilon^* \in [-1,1]} \hat{\lambda}_i\left(a, \varepsilon^*\right); \tag{6.37}$$

for $i = 1, 2, \ldots, l$.

Finally, all the fuzzy eigenvalue solutions ($\tilde{\lambda}_i$ for $i = 1, 2, \ldots, l$) of the FGEP (6.30) can be evaluated by varying its fuzzy parameter (a) from 0 to 1 and the fuzzy solution plots can also be constructed putting different values of a in the lower and upper bounds of the fuzzy parametric eigenvalue solutions (6.36) and (6.37), respectively.

Note 6.8

A similar procedure to that given in Section 6.7 (to solve the FGEP) may also be used to solve the FSEP $\tilde{S}\tilde{x} = \tilde{\lambda}\tilde{x}$. First, we have to transform the $l \times l$ fuzzy coefficient matrix $\tilde{S}$ into its fuzzy parametric representation ($\tilde{S}(a)$ for $a \in [0, 1]$) and then into the fuzzy-affine form ($\hat{S}(a, \varepsilon_{ij})$ for $i = 1, 2, \ldots, l$). Thus, the FSEP is converted into a fuzzy-affine SEP $\hat{S}(a, \varepsilon_{ij})\hat{x}(a, \varepsilon^*) = \hat{\lambda}(a, \varepsilon^*)\hat{x}(a, \varepsilon^*)$. Then, solving $\det\left(\hat{S}(a, \varepsilon_{ij}) - \hat{\lambda} I\right) = 0$, we get the required solution.

Example 6.9 Find the solution of a FGEP $\tilde{G}\tilde{x} = \tilde{\lambda}\tilde{H}\tilde{x}$, where one of its coefficient matrices is a 2×2 fuzzy matrix having elements in the form of TFNs and the other coefficient matrix is a 2×2 crisp matrix as given below.

$$\tilde{G} = \begin{bmatrix} (17800, 18000, 18200) & (-7300, -7200, -7100) \\ (-7300, -7200, -7100) & (7100, 7200, 7300) \end{bmatrix} \quad \text{and} \quad \tilde{H} = \begin{pmatrix} 3600 & 0 \\ 0 & 3600 \end{pmatrix}.$$

Table 6.5: Triangular fuzzy eigenvalue bounds ($\tilde{\lambda}$) of Example 6.9 for different values of a

a	First Eigenvalue ($\tilde{\lambda}_1$)		Second Eigenvalue ($\tilde{\lambda}_2$)	
	$\underline{\lambda_1}$	$\overline{\lambda_1}$	$\underline{\lambda_2}$	$\overline{\lambda_2}$
0	0.9443	1.0554	5.9724	6.0279
0.2	0.9555	1.0443	5.9779	6.0223
0.5	0.9722	1.0277	5.9861	6.0139
0.6	0.9778	1.0222	5.9889	6.0111
0.9	0.9944	1.0056	5.9972	6.0028
1	1.0000	1.0000	6.0000	6.0000

Also, plot all the fuzzy eigenvalue solutions of the given problem.

Solution: Here, the constituting elements of the given fuzzy coefficient matrix are in the form of TFNs. So, parameterizing the given fuzzy matrix $\tilde{G}$ by using a-cut approach, we may have

$$\tilde{K}(a) = \begin{pmatrix} [17800 + 200a, 18200 - 200a] & [-7300 + 100a, -7100 - 100a] \\ [-7300 + 100a, -7100 - 100a] & [7100 + 100a, 7300 - 100a] \end{pmatrix},$$

for $a \in [0, 1]$.

Further, the parametric fuzzy numbers are changed to their fuzzy-affine forms by adopting the procedure given in Section 6.7. Thus, the fuzzy-affine coefficient matrix is obtained as

$$\hat{K}\,(a, \varepsilon_i)_{i=1,\ldots,4} = \begin{bmatrix} 18000 + (200 - 200a)\varepsilon_1 & -7200 + (100 - 100a)\varepsilon_2 \\ -7200 + (100 - 100a)\varepsilon_3 & 7200 + (100 - 100a)\varepsilon_4 \end{bmatrix},$$

where $a \in [0, 1]$ and $\varepsilon_i \in [-1, 1]$ for $i = 1, 2, 3, 4$.

Thus, the given FGEP is converted to fuzzy-affine GEP $\hat{G}\hat{x} = \hat{\lambda}\hat{H}\hat{x}$, and it may be further solved to compute the fuzzy-affine eigenvalues having several parameters such as fuzzy parameters and noise symbols.

The fuzzy eigenvalues in the form of TFN are evaluated by adopting the proposed method given in Section 6.7. All the fuzzy eigenvalue plots are depicted in Figures 6.3 and 6.4.

Last, for some particular values of the fuzzy parameter viz. $a = 0, 0.2, 0.5, 0.6, 0.9$ and 1, the eigenvalues are listed in Table 6.5.

Example 6.10 Let us consider a five degrees-of-freedom spring-mass structural system as shown in Fig. 6.5 in an uncertain environment. Suppose all the structural parameters are considered in the form of TrFNs.

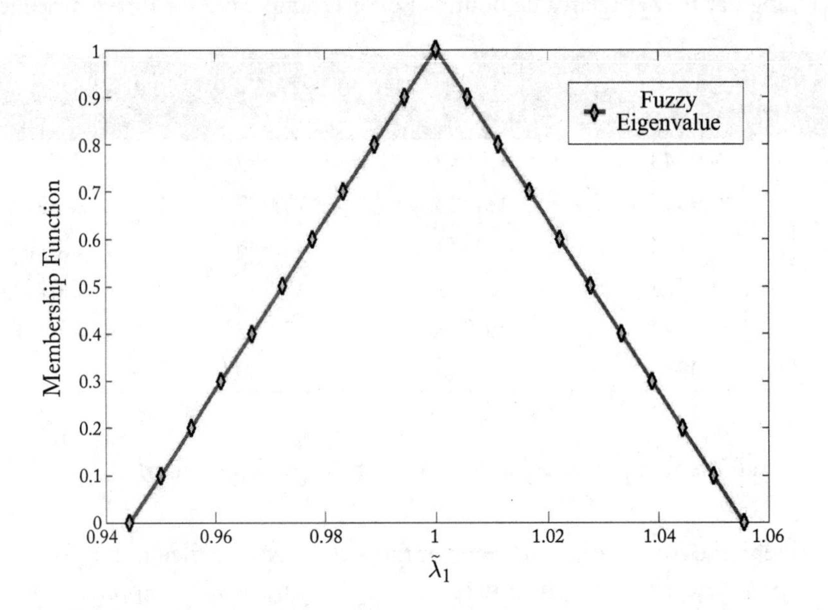

Figure 6.3: First fuzzy eigenvalue plot for Example 6.9.

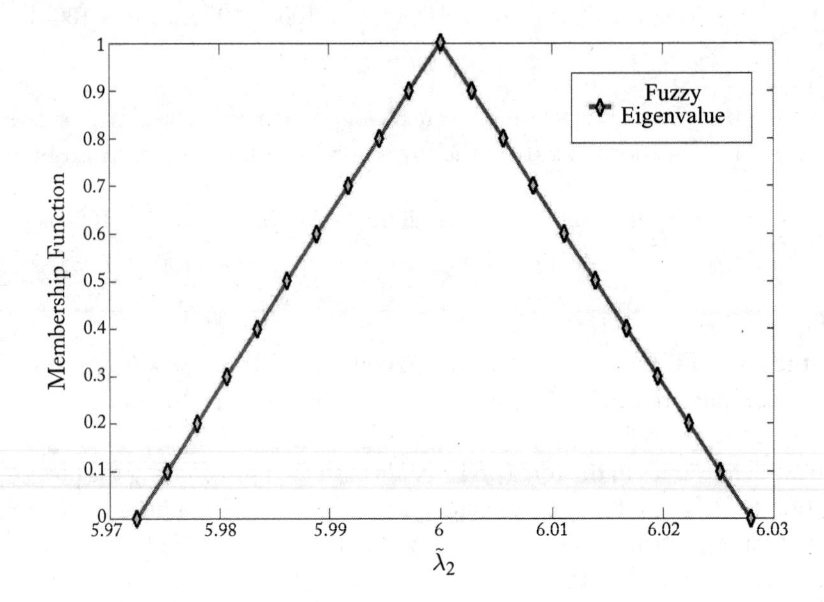

Figure 6.4: Second fuzzy eigenvalue plot for Example 6.9.

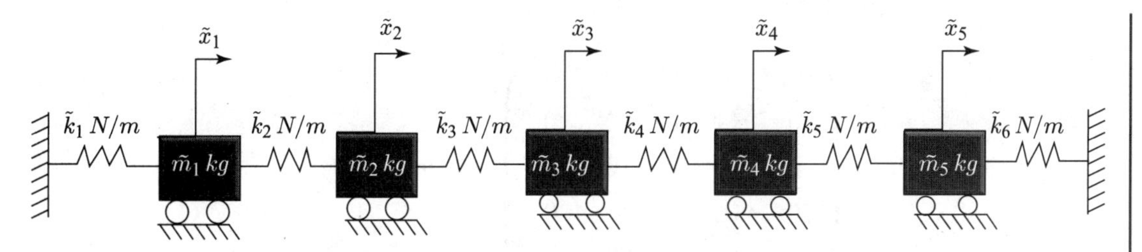

Figure 6.5: Spring-mass structural system of five degrees-of-freedom.

Here, the fuzzy stiffness and mass parameters in the form of TrFNs are given as

$$\begin{aligned}
\tilde{k}_1 &= (2000, 2020, 2080, 2100)\ (\text{N/m});\\
\tilde{k}_2 &= (1800, 1815, 1835, 1850)\ (\text{N/m});\\
\tilde{k}_3 &= (1600, 1612, 1618, 1630)\ (\text{N/m});\\
\tilde{k}_4 &= (1400, 1408, 1412, 1420)\ (\text{N/m});\\
\tilde{k}_5 &= (1200, 1203, 1207, 1210)\ (\text{N/m});\\
\tilde{k}_6 &= (1000, 1001, 1007, 1008)\ (\text{N/m}).
\end{aligned}$$

and

$$\begin{aligned}
\tilde{m}_1 &= (10, 10.5, 11.5, 12)\ (\text{kg});\\
\tilde{m}_2 &= (12, 12.4, 13.6, 14)\ (\text{kg});\\
\tilde{m}_3 &= (14, 14.2, 15.8, 16)\ (\text{kg});\\
\tilde{m}_4 &= (16, 16.8, 17.2, 18)\ (\text{kg});\\
\tilde{m}_5 &= (18, 18.6, 19.4, 20)\ (\text{kg}).
\end{aligned}$$

Evaluate the fuzzy eigenvalue bounds of the FGEP obtained during the dynamic analysis of the given spring-mass structural system. Also, plot all the fuzzy eigenvalue solutions.

Solution: From Eq. (6.3), the dynamic analysis of the five degrees-of-freedom spring-mass structural system having fuzzy parameters leads to a FGEP $\tilde{K}\tilde{x} = \tilde{\lambda}\tilde{M}\tilde{x}$, where the fuzzy stiffness and mass matrices (from Eq. (6.2)) may be written as

$$\tilde{K} = \begin{pmatrix} \tilde{k}_1 + \tilde{k}_2 & -\tilde{k}_2 & 0 & 0 & 0 \\ -\tilde{k}_2 & \tilde{k}_2 + \tilde{k}_3 & -\tilde{k}_3 & 0 & 0 \\ 0 & -\tilde{k}_3 & \tilde{k}_3 + \tilde{k}_4 & -\tilde{k}_4 & 0 \\ 0 & 0 & -\tilde{k}_4 & \tilde{k}_4 + \tilde{k}_5 & -\tilde{k}_5 \\ 0 & 0 & 0 & -\tilde{k}_5 & \tilde{k}_5 + \tilde{k}_6 \end{pmatrix}$$

and

$$\tilde{M} = \begin{pmatrix} \tilde{m}_1 & 0 & 0 & 0 & 0 \\ 0 & \tilde{m}_2 & 0 & 0 & 0 \\ 0 & 0 & \tilde{m}_3 & 0 & 0 \\ 0 & 0 & 0 & \tilde{m}_4 & 0 \\ 0 & 0 & 0 & 0 & \tilde{m}_5 \end{pmatrix}.$$

Here, the fuzzy stiffness and mass parameters are given. Thus, the fuzzy parametric forms of the given stiffness and mass parameters are found as

$$\tilde{k}_1(a) = [2000 + 20a, 2100 - 20a]\ \text{N/m}; \quad \tilde{k}_2 = [1800 + 15a, 1850 - 15a]\ \text{N/m};$$
$$\tilde{k}_3(a) = [1600 + 12a, 1630 - 12a]\ \text{N/m}; \quad \tilde{k}_4(a) = [1400 + 8a, 1420 - 8a]\ \text{N/m};$$
$$\tilde{k}_5(a) = [1200 + 3a, 1210 - 3a]\ \text{N/m}; \quad \tilde{k}_6(a) = [1000 + a, 1008 - a]\ \text{N/m}$$

and

$$\tilde{m}_1(a) = [10 + 0.5a, 12 - 0.5a]\ \text{kg}; \quad \tilde{m}_2(a) = [12 + 0.4a, 14 - 0.4a]\ \text{kg};$$
$$\tilde{m}_3(a) = [14 + 0.2a, 16 - 0.2a]\ \text{kg}; \quad \tilde{m}_4(a) = [16 + 0.8a, 18 - 0.8a]\ \text{kg};$$
$$\tilde{m}_5(a) = [18 + 0.6a, 20 - 0.6a]\ \text{kg},$$

for $a \in [0, 1]$.

Further, the parametric fuzzy numbers are changed to their fuzzy-affine forms by adopting the procedure given in Section 6.7 as follows:

$$\hat{k}_1(a, \varepsilon_1) = 2050 + (50 - 20a)\varepsilon_1\ \text{N/m}; \ \hat{k}_2(a, \varepsilon_2) = 1825 + (25 - 15a)\varepsilon_2\ \text{N/m};$$
$$\hat{k}_3(a, \varepsilon_3) = 1615 + (15 - 12a)\varepsilon_3\ \text{N/m}; \ \hat{k}_4(a, \varepsilon_4) = 1410 + (10 - 8a)\varepsilon_4\ \text{N/m};$$
$$\hat{k}_5(a, \varepsilon_5) = 1205 + (5 - 3a)\varepsilon_5\ \text{N/m}; \ \hat{k}_6(a, \varepsilon_6) = 1004 + (4 - a)\varepsilon_6\ \text{N/m}.$$

and

$$\hat{m}_1(a, \varepsilon_7) = 11 + (1 - 0.5a)\varepsilon_7\ \text{kg}; \ \hat{m}_2(a, \varepsilon_8) = 13 + (1 - 0.4a)\varepsilon_8\ \text{kg};$$
$$\hat{m}_3(a, \varepsilon_9) = 15 + (1 - 0.2a)\varepsilon_9\ \text{kg}; \ \hat{m}_4(a, \varepsilon_{10}) = 17 + (1 - 0.8a)\varepsilon_{10}\ \text{kg};$$
$$\hat{m}_5(a, \varepsilon_{11}) = 19 + (1 - 0.6a)\varepsilon_{11}\ \text{kg}, \text{ where } a \in [0, 1] \text{ and } \varepsilon_i \in [-1, 1] \text{ for } i = 1, \ldots, 11.$$

Hence, the given FGEP $\tilde{K}\tilde{x} = \tilde{\lambda}\tilde{M}\tilde{x}$ is converted to fuzzy-affine GEP $\hat{K}\hat{x} = \hat{\lambda}\hat{M}\hat{x}$, where the fuzzy-affine coefficient matrices (that is, fuzzy-affine stiffness and mass matrices) are obtained as

$$\hat{K}(a,\varepsilon_i)_{i=1,\ldots,6} =$$
$$\begin{pmatrix} \hat{k}_1(a,\varepsilon_1)+\hat{k}_2(a,\varepsilon_2) & -\hat{k}_2(a,\varepsilon_2) & 0 & 0 & 0 \\ -\hat{k}_2(a,\varepsilon_2) & \hat{k}_2(a,\varepsilon_2)+\hat{k}_3(a,\varepsilon_3) & -\hat{k}_3(a,\varepsilon_3) & 0 & 0 \\ 0 & -\hat{k}_3(a,\varepsilon_3) & \hat{k}_3(a,\varepsilon_3)+\hat{k}_4(a,\varepsilon_4) & -\hat{k}_4(a,\varepsilon_4) & 0 \\ 0 & 0 & -\hat{k}_4(a,\varepsilon_4) & \hat{k}_4(a,\varepsilon_4)+\hat{k}_5(a,\varepsilon_5) & -\hat{k}_5(a,\varepsilon_5) \\ 0 & 0 & 0 & -\hat{k}_5(a,\varepsilon_5) & \hat{k}_5(a,\varepsilon_5)+\hat{k}_6(a,\varepsilon_6) \end{pmatrix}$$

and

$$\hat{M}(a,\varepsilon_i)_{i=7,\ldots,11} = \begin{pmatrix} \hat{m}_1(a,\varepsilon_7) & 0 & 0 & 0 & 0 \\ 0 & \hat{m}_2(a,\varepsilon_8) & 0 & 0 & 0 \\ 0 & 0 & \hat{m}_3(a,\varepsilon_9) & 0 & 0 \\ 0 & 0 & 0 & \hat{m}_4(a,\varepsilon_{10}) & 0 \\ 0 & 0 & 0 & 0 & \hat{m}_5(a,\varepsilon_{11}) \end{pmatrix}.$$

That is,

$$\hat{K}(a,\varepsilon_i)_{i=1,\ldots,6} = \begin{pmatrix} 3875+(50-20a)\varepsilon_1+(25-15a)\varepsilon_2 & -1825-(25-15a)\varepsilon_2 & 0 & 0 & 0 \\ -1825-(25-15a)\varepsilon_2 & 3440+(25-15a)\varepsilon_2+(15-12a)\varepsilon_3 & -1615-(15-12a)\varepsilon_3 & 0 & 0 \\ 0 & -1615-(15-12a)\varepsilon_3 & 3025+(15-12a)\varepsilon_3+(10-8a)\varepsilon_4 & -1410-(10-8a)\varepsilon_4 & 0 \\ 0 & 0 & -1410-(10-8a)\varepsilon_4 & 2615+(10-8a)\varepsilon_4+(5-3a)\varepsilon_5 & -1205-(5-3a)\varepsilon_5 \\ 0 & 0 & 0 & -1205-(5-3a)\varepsilon_5 & 2209+(5-3a)\varepsilon_5+(4-a)\varepsilon_6 \end{pmatrix}$$

and

$$\hat{M}(a,\varepsilon_i)_{i=7,\ldots,11} = \begin{pmatrix} 11+(1-0.5a)\varepsilon_7 & 0 & 0 & 0 & 0 \\ 0 & 13+(1-0.4a)\varepsilon_8 & 0 & 0 & 0 \\ 0 & 0 & 15+(1-0.2a)\varepsilon_9 & 0 & 0 \\ 0 & 0 & 0 & 17+(1-0.8a)\varepsilon_{10} & 0 \\ 0 & 0 & 0 & 0 & 19+(1-0.6a)\varepsilon_{11} \end{pmatrix},$$

where $a \in [0,1]$ and $\varepsilon_i \in [-1,1]$ for $i = 1,\ldots,11$.

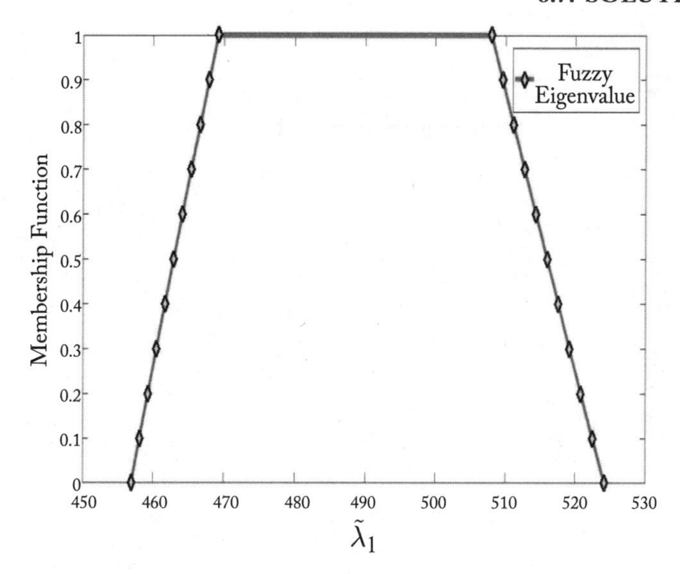

Figure 6.6: First eigenvalue plot for Example 6.10.

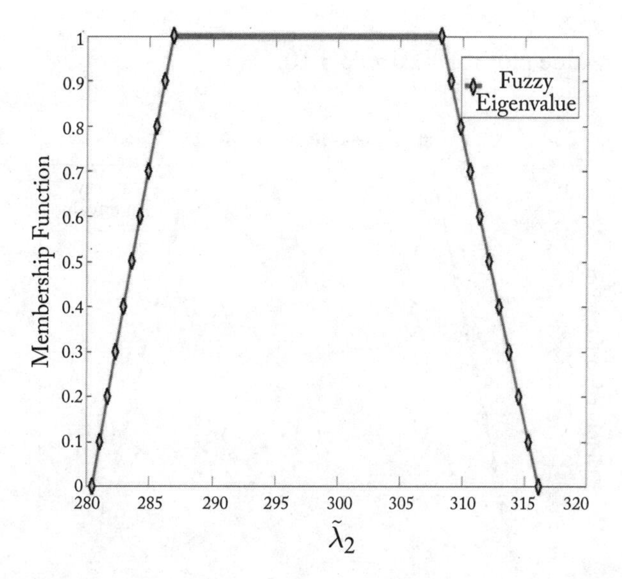

Figure 6.7: Second eigenvalue plot for Example 6.10.

Then, all the fuzzy eigenvalues in the form of TrFN are evaluated by adopting the proposed method given in Section 6.7. The fuzzy eigenvalue plots are depicted in Figures 6.6–6.10.

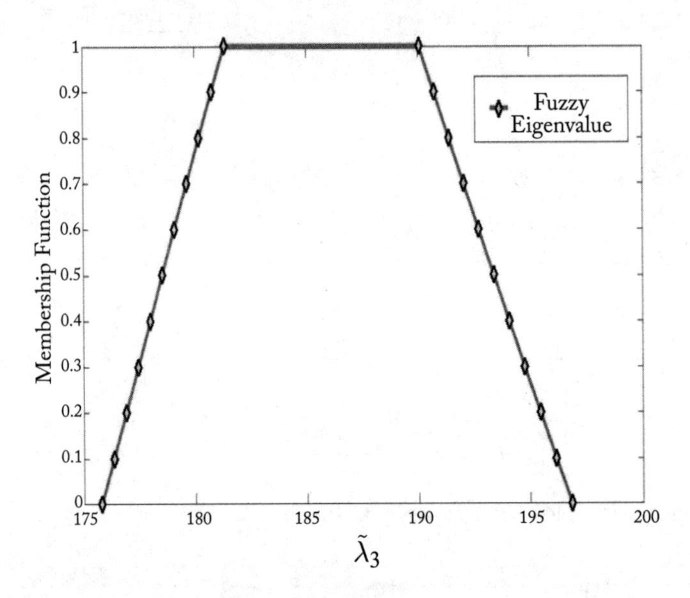

Figure 6.8: Third eigenvalue plot for Example 6.10.

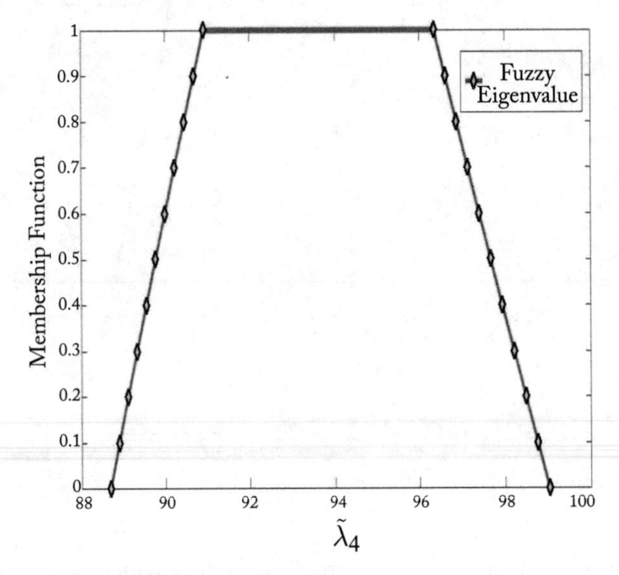

Figure 6.9: Fourth eigenvalue plot for Example 6.10.

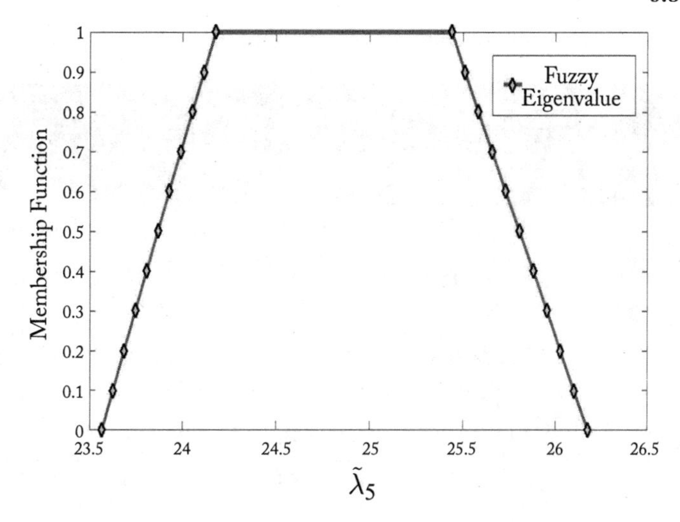

Figure 6.10: Fifth eigenvalue plot for Example 6.10.

Finally, in Table 6.6, all the fuzzy eigenvalue solutions of Example 6.10 are listed for some particular values of the fuzzy parameter viz. $a = 0, 0.1, 0.4, 0.5, 0.8$, and 1.

6.8 EXERCISES

6.1. Compute all the eigenvalue solutions of the crisp standard eigenvalue problems $S_1 x = \lambda x$ and $S_2 u = \mu u$, where

$$S_1 = \begin{pmatrix} -3 & -8.5 & 14.5 & 4.8 & -1.1 \\ 17.5 & 17.5 & 1.5 & 4.5 & 10.5 \\ 17.1 & -3 & 2 & -12.5 & 6.2 \\ 18.5 & 2.5 & 18.5 & 5.5 & 6.5 \\ 13.5 & 18.5 & 9.5 & -17.5 & 10.5 \end{pmatrix}$$

and

$$S_2 = \begin{pmatrix} 3000 & -2000 & 0 & 0 \\ -2000 & 5000 & -3000 & 0 \\ 0 & -3000 & 7000 & -4000 \\ 0 & 0 & -4000 & 9000 \end{pmatrix}.$$

6.2. Evaluate the lower and upper bounds of the interval eigenvalue solutions for the following interval standard eigenvalue problems.

$$[S_1][x] = [\lambda][x] \quad \text{and} \quad [S_2][u] = [\mu][u],$$

Table 6.6: Trapezoidal fuzzy eigenvalue bounds of Example 6.10 for different values of a

a	Lower and Upper Bounds	λ_1	λ_2	λ_3	λ_4	λ_5
0	$\underline{\lambda}$	456.8997	280.4391	175.7996	88.6697	23.5660
	$\overline{\lambda}$	524.1465	316.1529	196.8291	99.0332	26.1796
0.1	$\underline{\lambda}$	458.0849	281.0606	176.3378	88.8865	23.6256
	$\overline{\lambda}$	522.4658	315.3404	196.1294	98.7558	26.1035
0.4	$\underline{\lambda}$	461.6980	282.9542	177.9735	89.5451	23.8064
	$\overline{\lambda}$	517.5181	312.9429	194.0622	97.9352	25.8780
0.5	$\underline{\lambda}$	462.9220	283.5954	178.5260	89.7674	23.8673
	$\overline{\lambda}$	515.8996	312.1567	193.3837	97.6654	25.8038
0.8	$\underline{\lambda}$	466.6545	285.5496	180.2055	90.4429	24.0523
	$\overline{\lambda}$	511.1332	309.8355	191.3787	96.8667	25.5841
1	$\underline{\lambda}$	469.1946	286.8788	181.3439	90.9004	24.1774
	$\overline{\lambda}$	508.0278	308.3185	190.0671	96.3431	25.4399

where

$$[S_1] = \begin{pmatrix} [-5,-4] & [-9,-8] & [14,15] & [4.6,5] & [-1.2,-1] \\ [17,18] & [17,18] & [1,2] & [4,5] & [10,11] \\ [17,17.2] & [-3.5,-2.7] & [1.9,2.1] & [-13,-12] & [6,6.4] \\ [18,19] & [2,3] & [18,19] & [5,6] & [6,7] \\ [13,14] & [18,19] & [9,10] & [-18,-17] & [10,11] \end{pmatrix}$$

and

$$[S_2] = \begin{pmatrix} [2975,3025] & [-2015,-1985] & 0 & 0 \\ [-2015,-1985] & [4965,5035] & [-3020,-2980] & 0 \\ 0 & [-3020,-2980] & [6955,7045] & [-4025,-3975] \\ 0 & 0 & [-4025,-3975] & [8945,9055] \end{pmatrix}.$$

Also, verify whether the interval solutions contain the crisp solutions given in Exercise 6.1 or not.

6.3. Determine the triangular fuzzy solution plots of the fuzzy generalized eigenvalue problem $\tilde{G}\tilde{x} = \tilde{\lambda}\tilde{H}\tilde{x}$, where the two coefficient matrices are 5×5 fuzzy matrices whose elements are in the form of TFN given as follows:

$$\tilde{G} = \begin{bmatrix} (3800,3835,3870) & (-1850,-1825,-1800) & 0 & 0 & 0 \\ (-1850,-1825,-1800) & (3400,3440,3480) & (-1630,-1615,-1600) & 0 & 0 \\ 0 & (-1630,-1615,-1600) & (3000,3025,3050) & (-1420,-1410,-1400) & 0 \\ 0 & 0 & (-1420,-1410,-1400) & (2600,2615,2630) & (-1210,-1205,-1200) \\ 0 & 0 & 0 & (-1210,-1205,-1200) & (1200,1205,1210) \end{bmatrix}$$

and

$$\tilde{H} = \begin{bmatrix} (29.5,30,30.5) & 0 & 0 & 0 & 0 \\ 0 & (26,27,28) & 0 & 0 & 0 \\ 0 & 0 & (26.5,27,27.5) & 0 & 0 \\ 0 & 0 & 0 & (23.5,25,26.5) & 0 \\ 0 & 0 & 0 & 0 & (17,18,19) \end{bmatrix}.$$

6.9 REFERENCES

[1] Chakraverty, S. and Behera, D., 2014. Parameter identification of multistorey frame structure from uncertain dynamic data. *Strojniški Vestnik-Journal of Mechanical Engineering*, 60(5):331–338. DOI: 10.5545/sv-jme.2014.1832. 98

[2] Chakraverty, S. and Behera, D., 2017. Uncertain static and dynamic analysis of imprecisely defined structural systems. In *Fuzzy Systems: Concepts, Methodologies, Tools, and Applications*, pages 1–30, IGI Global. DOI: 10.4018/978-1-5225-1908-9.ch001. 98

[3] Hladík, M., 2013. Bounds on eigenvalues of real and complex interval matrices. *Applied Mathematics and Computation*, 219(10):5584–5591. DOI: 10.1016/j.amc.2012.11.075.

[4] Hladik, M., Daney, D., and Tsigaridas, E., 2011. A filtering method for the interval eigenvalue problem. *Applied Mathematics and Computation*, 217(12):5236–5242. DOI: 10.1016/j.amc.2010.09.066.

[5] Leng, H., 2014. Real eigenvalue bounds of standard and generalized real interval eigenvalue problems. *Applied Mathematics and Computation*, 232:164–171. DOI: 10.1016/j.amc.2014.01.070.

[6] Leng, H. and He, Z., 2007. Computing eigenvalue bounds of structures with uncertain-but-non-random parameters by a method based on perturbation theory. *Communications in Numerical Methods in Engineering*, 23(11):973–982. DOI: 10.1002/cnm.936.

[7] Leng, H., He, Z., and Yuan, Q., 2008. Computing bounds to real eigenvalues of real-interval matrices. *International Journal for Numerical Methods in Engineering*, 74(4):523–530. DOI: 10.1002/nme.2179.

[8] Mahato, N. R. and Chakraverty, S., 2016a. Filtering algorithm for real eigenvalue bounds of interval and fuzzy generalized eigenvalue problems. *ASCE-ASME Journal of Risk and Uncertainty in Engineering Systems, Part B: Mechanical Engineering*, 2(4):044502. DOI: 10.1115/1.4032958.

[9] Mahato, N. R. and Chakraverty, S., 2016b. Filtering algorithm for eigenvalue bounds of fuzzy symmetric matrices. *Engineering Computations*, 33(3):855–875. DOI: 10.1108/ec-12-2014-0255. 106, 107, 108

[10] Qiu, Z., Chen, S., and Elishakoff, I., 1996. Bounds of eigenvalues for structures with an interval description of uncertain-but-non-random parameters. *Chaos, Solitons and Fractals*, 7(3):425–434. DOI: 10.1016/0960-0779(95)00065-8.

[11] Qiu, Z., Chen, S., and Jia, H., 1995. The Rayleigh quotient iteration method for computing eigenvalue bounds of structures with bounded uncertain parameters. *Computers and Structures*, 55(2):221–227. DOI: 10.1016/0045-7949(94)00444-8.

[12] Qiu, Z., Wang, X., and Friswell, M. I., 2005. Eigenvalue bounds of structures with uncertain-but-bounded parameters. *Journal of Sound and Vibration*, 282(1–2):297–312. DOI: 10.1016/j.jsv.2004.02.051.

[13] Sim, J., Qiu, Z., and Wang, X., 2007. Modal analysis of structures with uncertain-but-bounded parameters via interval analysis. *Journal of Sound and Vibration*, 303(1–2):29–45. DOI: 10.1016/j.jsv.2006.11.038.

[14] Xia, Y. and Friswell, M., 2014. Efficient solution of the fuzzy eigenvalue problem in structural dynamics. *Engineering Computations*, 31(5):864–878. DOI: 10.1108/ec-02-2013-0052.

CHAPTER 7

Uncertain Nonlinear Dynamic Problems

Nonlinear dynamic problems from various fields of science and engineering lead to nonlinear eigenvalue problems. In this chapter, we focus on the solutions of nonlinear eigenvalue problems with uncertainty. A nonlinear eigenvalue problem is a generalization of a linear eigenvalue problem viz. standard eigenvalue problem or generalized eigenvalue problem to the equations that depend nonlinearly on the eigenvalues. Mathematically, a nonlinear eigenvalue problem is generally described by an equation of the form $M(\lambda)x = 0$, for all λ, and contains two unknowns viz. the eigenvalue parameter (λ) and the "nontrivial" vector(s) (x) (known as eigenvector) corresponding to it.

For simplicity and easy computations, all the involved parameters and variables of a physical system are usually considered deterministic or exact. But as a practical matter, due to the uncertain environment, one may have imprecise, incomplete, insufficient, or vague information about the parameters because of several errors. Traditionally, such uncertainty or vagueness may be modeled through a probabilistic approach. But a large amount of data is required for the traditional probabilistic approach. Without a sufficient amount of experimental data, the probabilistic methods may not deliver reliable results at the required precision. Therefore, intervals and/or fuzzy numbers may become efficient tools to handle uncertain and vague parameters when there is an insufficient amount of data available. In this regard, uncertain nonlinear dynamic problems may be modeled through an interval nonlinear eigenvalue problem (INEP) ($[M]([\lambda])[x] = 0$) and/or fuzzy nonlinear eigenvalue problem (FNEP) $\left(\tilde{M}(\tilde{\lambda})\tilde{x} = 0\right)$.

The nonlinear eigenvalue problem has a wide variety of applications in the dynamic analysis of various science and engineering problems viz. structural mechanics, electrical circuit simulations, micro-electronic mechanical systems, acoustic systems, signal processing, and fluid mechanics. For instance, the nonlinear eigenvalue problem plays a very important role in the application of structural dynamics. In general, dynamic analysis of structural problems of the damped spring-mass system gets converted into a nonlinear eigenvalue problem (particularly, quadratic eigenvalue problem (QEP)).

The equation of motion for the structural ambient vibrational problems of the system given in Fig. 7.1 may be written as

$$M\ddot{u}(t) + C\dot{u}(t) + Ku(t) = 0, \tag{7.1}$$

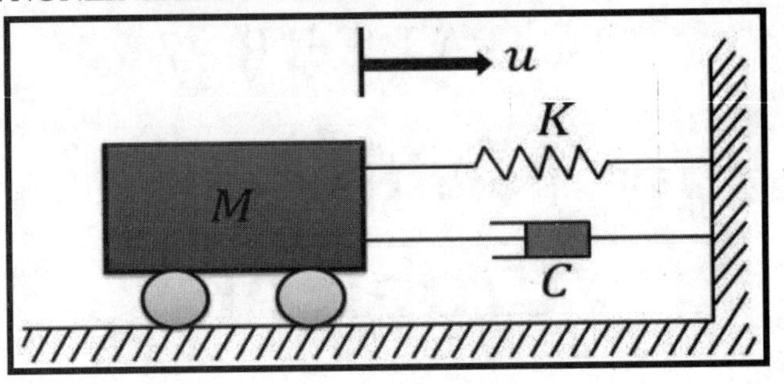

Figure 7.1: One degree-of-freedom damped spring-mass structural system.

where M is the mass matrix, C is the damping matrix, and K is the stiffness matrix. Substituting $u = xe^{i\omega \cdot t}$ in the governing equation of motion given in Eq. (7.1) may be reduced to a QEP as

$$\left(K + \lambda C + \lambda^2 M\right) x = 0, \tag{7.2}$$

where $\lambda = i\omega$ stands for the eigenvalue and x is the corresponding eigenvector of the above problem.

Moreover, the study for corner singularities in case of anisotropic elastic materials [Mehrmann and Watkins (2002) [14]; Apel, Mehrmann, and Watkins (2002) [1]] leads to the quadratic eigenvalue problem of the following form:

$$Kx + \lambda Gx + \lambda^2 Mx = 0, \tag{7.3}$$

where $K = K^T$, $G = -G^T$, and $M = M^T$.

Therefore, the nonlinear eigenvalue problem plays a vital role in various application problems. Keeping this in view, evaluating the solution of the nonlinear eigenvalue problem becomes a very important and challenging task for many researchers.

7.1 CRISP NONLINEAR EIGENVALUE PROBLEM (CNEP)

Suppose $M(\lambda)$ is a nonlinear matrix-valued function such that $M : \Delta \to C^{k \times k}$, where $\Delta \subseteq C$ is an open set. Then, the nonlinear eigenvalue problem (NEP) is finding the ordered pairs $(x, \lambda) \in C^k \times \Delta$ (for $x \neq 0$) such that

$$M(\lambda)x = 0. \tag{7.4}$$

For such an ordered pair (x, λ), λ is known as the eigenvalue and x is the corresponding eigenvector of λ. Here, k is the degree of the NEP. When the parameters of the NEP are crisp (or exact) numbers, it may be referred to as a crisp nonlinear eigenvalue problem (CNEP). Thus, the general form of the CNEP of degree k may be defined as

$$M(\lambda)x = \sum_{j=0}^{k} \lambda^j M_j x = \left(M_k \lambda^k + M_{k-1} \lambda^{k-1} + \cdots + M_1 \lambda + M_0\right) x = 0, \tag{7.5}$$

where all the coefficient matrices M_j for $j = 0, 1, \ldots, k$ are crisp positive definite square matrices. Now, depending upon the values of k (that is, the degree of the eigenvalue λ in the nonlinear polynomial function $M(\lambda)$), CNEPs may be classified as follows.

7.1.1 CRISP QUADRATIC EIGENVALUE PROBLEM (CQEP)

The CNEP $M(\lambda)x = 0$ may be referred to as a crisp quadratic eigenvalue problem (CQEP) when the degree of λ in the nonlinear matrix-valued function $M(\lambda)$ is two (that is $k = 2$). In that case, the quadratic eigenvalue problem may be defined as

$$M(\lambda)x = \left(M_2\lambda^2 + M_1\lambda + M_0\right)x = 0, \tag{7.6}$$

where the coefficient matrices M_2, M_1, and M_0 are positive definite square matrices having crisp entries.

7.1.2 CRISP CUBIC EIGENVALUE PROBLEM (CCEP)

The nonlinear matrix-valued function $M(\lambda)$ having degree of three may be considered as a crisp cubic eigenvalue problem (CCEP) and is defined as

$$M(\lambda)x = \left(M_3\lambda^3 + M_2\lambda^2 + M_1\lambda + M_0\right)x = 0, \tag{7.7}$$

where the coefficient matrices M_j for $j = 0, 1, 2, 3$ are considered to be crisp positive definite square matrices.

Note 7.1

1. For a NEP (in Eq. (7.5)) $k \geq 2$. For $k = 1$, it may be referred to as a linear eigenvalue problem (described in Chapter 6).
2. When $k = 1$ and $M_0 = G$ and $M_1 = -H$, then the nonlinear matrix function becomes $M(\lambda) = G - \lambda H$ and is known as a generalized eigenvalue problem (GEP).
3. When $k = 1$ and $M_0 = S$ and $M_1 = -I$ ($I \to$ identity matrix), then the nonlinear matrix function becomes $M(\lambda) = S - \lambda I$ and is known as a standard eigenvalue problem (SEP).

7.2 LINEARIZATION OF CNEP

The linearization technique plays a major role in determining the solutions of a CNEP. After linearizing a CNEP, it may be transformed into a crisp generalized eigenvalue problem (CGEP). Works related to the linearization technique of the NEP having crisp (or exact) parameters have been studied by Tisseur and Meerbergen (2001) [21], Mehrmann and Watkins (2002) [14], and others.

Let $M(\lambda)x = 0$ be a CNEP of degree k. Then, its general form (as given in Eq. (7.5)) may be expressed as

$$M(\lambda)x = \sum_{j=0}^{k} \lambda^j M_j x = \left(M_k\lambda^k + M_{k-1}\lambda^{k-1} + \cdots + M_1\lambda + M_0\right)x = 0,$$

$$\Rightarrow M_k\lambda^k x + M_{k-1}\lambda^{k-1}x + \cdots + M_1\lambda x + M_0 x = 0. \tag{7.8}$$

We define new variables $x_1, x_2, \ldots, x_n$ in the following manner.

$$x := x_1,\ x_2 := \lambda x_1 = \lambda x,\ x_3 := \lambda x_2 = \lambda^2 x,\ \ldots,\ x_k := \lambda x_{k-1} = \lambda^k x. \tag{7.9}$$

Substituting these new variables (from Eq. (7.9)) in the Eq. (7.8), we may have the following linear eigenvalue problem.

$$M_k\lambda x_k + M_{k-1}\lambda x_{k-1} + \cdots + M_2\lambda x_2 + M_1\lambda x_1 + M_0 x_1 = 0. \tag{7.10}$$

The above Eqs. (7.9) and (7.10) may form a system of linear eigenvalue problem as follows:

$$\left.\begin{aligned} -M_0x_1 &= \lambda(M_1x_1 + M_2x_2 + \cdots + M_{k-1}x_{k-1} + M_kx_k) \\ x_2 &= \lambda x_1 \\ &\ \vdots \\ x_k &= \lambda x_{k-1} \end{aligned}\right\}. \tag{7.11}$$

Further, the above system can be represented in the matrix form given below.

$$\left(\begin{array}{c|ccc} -M_0 & 0 & \cdots & 0 \\ \hline 0 & I & & 0 \\ \vdots & \vdots & \ddots & \vdots \\ 0 & 0 & \cdots & I \end{array}\right) \cdot \left(\begin{array}{c} x_1 \\ \hline x_2 \\ \vdots \\ x_k \end{array}\right) = \lambda \cdot \left(\begin{array}{ccc|c} M_1 & \cdots & M_{k-1} & M_k \\ \hline I & & 0 & 0 \\ \vdots & \ddots & \vdots & \vdots \\ 0 & \cdots & I & 0 \end{array}\right) \cdot \left(\begin{array}{c} x_1 \\ \vdots \\ x_{k-1} \\ \hline x_k \end{array}\right). \tag{7.12}$$

The system (7.12) yields a generalized eigenvalue problem (GEP) $Gx^{\#} = \lambda Hx^{\#}$, where $(G - \lambda H) \in \mathbb{C}^{kl,kl}$ and

$$G = \left(\begin{array}{c|ccc} -M_0 & 0 & \cdots & 0 \\ \hline 0 & I & & 0 \\ \vdots & \vdots & \ddots & \vdots \\ 0 & 0 & \cdots & I \end{array}\right)_{kl\times kl} \quad \text{and} \quad H = \left(\begin{array}{ccc|c} M_1 & \cdots & M_{k-1} & M_k \\ \hline I & & 0 & 0 \\ \vdots & \ddots & \vdots & \vdots \\ 0 & \cdots & I & 0 \end{array}\right)_{kl\times kl}. \tag{7.13}$$

Here, "l" is the dimension of all the coefficient matrices M_j for $j = 0, 1, \ldots, k$ and $x^{\#} = \left[{x_1}^T, {x_2}^T, \ldots, {x_k}^T\right]^T = \left[x^T, \lambda x^T, \ldots, \lambda^{k-1}x^T\right]^T$ is the newly generated eigenvector.

Note 7.2

1. The NEP $M(\lambda)x = 0$ and the newly formed GEP $Gx^{\#} = \lambda H x^{\#}$ have the same eigenvalues.

2. Since the coefficient matrices of the GEP (7.12) are of order "kl," then there exists "kl" number of eigenvalues of the NEP $M(\lambda)x = 0$ of degree k.

7.3 SOLUTION OF CNEP

Suppose $M(\lambda)x = 0$ is a CNEP having degree k and $l \times l$ coefficient matrices. Then, as given in Section 7.2, the CNEP may be linearized into a CGEP as $Gx^{\#} = \lambda H x^{\#}$ (Eq. (7.13)). Further, there exist various well-known methods to solve CGEP. By utilizing any of these methods, the newly formed CGEP may be solved easily.

Example 7.3 Determine all the eigenvalue solutions of the CCEP $M(\lambda)x = (M_3\lambda^3 + M_2\lambda^2 + M_1\lambda + M_0)x = 0$, where the coefficient matrices are taken as two-dimensional crisp square matrices given as follows:

$$M_3 = \begin{pmatrix} 0 & 0 \\ 0 & 0 \end{pmatrix}, \quad M_2 = \begin{pmatrix} 1 & 0 \\ 0 & 1 \end{pmatrix}, \quad M_1 = \begin{pmatrix} 1 & 2 \\ 2 & 1 \end{pmatrix}, \quad \text{and } M_0 = \begin{pmatrix} 1 & 2 \\ -1 & -1 \end{pmatrix}.$$

Solution: As given in Section 7.2, the given CCEP may be transformed (by using linearization technique) into a GEP $Gx^{\#} = \lambda H x^{\#}$, where

$$G = \left(\begin{array}{c|cc} -M_0 & 0 & 0 \\ \hline 0 & I & 0 \\ 0 & 0 & I \end{array}\right)_{6\times 6} \quad \text{and} \quad H = \left(\begin{array}{cc|c} M_1 & M_2 & M_3 \\ \hline I & 0 & 0 \\ 0 & I & 0 \end{array}\right)_{6\times 6}.$$

That is,

$$G = \left(\begin{array}{cc|cccc} -1 & -2 & 0 & 0 & 0 & 0 \\ 1 & 1 & 0 & 0 & 0 & 0 \\ \hline 0 & 0 & 1 & 0 & 0 & 0 \\ 0 & 0 & 0 & 1 & 0 & 0 \\ 0 & 0 & 0 & 0 & 1 & 0 \\ 0 & 0 & 0 & 0 & 0 & 1 \end{array}\right)_{6\times 6} \quad \text{and} \quad H = \left(\begin{array}{cccc|cc} 1 & 2 & 1 & 0 & 1 & 0 \\ 2 & 1 & 0 & 1 & 0 & 1 \\ \hline 1 & 0 & 0 & 0 & 0 & 0 \\ 0 & 1 & 0 & 0 & 0 & 0 \\ 0 & 0 & 1 & 0 & 0 & 0 \\ 0 & 0 & 0 & 1 & 0 & 0 \end{array}\right)_{6\times 6}.$$

Table 7.1: Crisp eigenvalue solutions (λ) of Example 7.3

i	Eigenvalues (λ_i)
1	0.7023
2	0.3792
3	-1.0000
4	-1.2709
5	-0.4053 + 1.6704i
6	-0.4053 - 1.6704i

Here, the degree of the CNEP is three and the coefficient matrices are of dimension 2×2. Therefore, there are $3 \times 2 = 6$ eigenvalues of the given problem. To compute these eigenvalues, we may solve the above GEP by using any well-known method. All the solutions (obtained by using MATLAB) are listed in Table 7.1.

Example 7.4 Find all the eigenvalues of the CQEP $M(\lambda)x = (M_2\lambda^2 + M_1\lambda + M_0)x = 0$ [Fazeli and Rabiei (2016) [5]], where the coefficient matrices are taken as 4×4 crisp matrices given as follows:

$$M_2 = \begin{pmatrix} 1 & 0.17 & -0.25 & 0.54 \\ 0.47 & 1 & 0.67 & -0.32 \\ -0.11 & 0.35 & 1 & -0.74 \\ 0.55 & 0.43 & 0.36 & 1 \end{pmatrix}, \quad M_1 = \begin{pmatrix} 0.22 & 0.02 & 0.12 & 0.14 \\ 0.02 & 0.14 & 0.04 & -0.06 \\ 0.12 & 0.04 & 0.28 & 0.08 \\ 0.14 & -0.06 & 0.08 & 0.26 \end{pmatrix}$$

and

$$M_0 = \begin{pmatrix} -3.0475 & -2.1879 & -1.9449 & -2.8242 \\ -2.6500 & -2.4724 & -2.3515 & -2.1053 \\ -0.7456 & -0.6423 & -1.3117 & -0.1852 \\ -4.0500 & -3.0631 & -2.8121 & -3.7794 \end{pmatrix}.$$

Solution: The given CNEP is of degree two, having 4×4 coefficient matrices. Thus, the problem has $2 \times 4 = 8$ eigenvalues. As given in Example 7.3, this CQEP may also be converted into a CGEP and is solved to find all the eigenvalues which are incorporated in Table 7.2.

Table 7.2: Crisp eigenvalue solutions (λ) of Example 7.4

i	Eigenvalues (λ_i)
1	2.3227
2	0.7967
3	0.2422
4	0.6382
5	-0.3776
6	-0.8394
7	-1.2234
8	-2.6353

7.4 INTERVAL NONLINEAR EIGENVALUE PROBLEM (INEP)

When the parameters of the NEP are uncertain or vague (particularly taken in the form of closed intervals), it may be addressed as an interval nonlinear eigenvalue problem (INEP). The INEP may be defined to find the interval scalars ($[\lambda]$) and nonzero interval vectors ($[x]$) such that

$$[M]([\lambda])[x] = 0, \tag{7.14}$$

where $[M]([\lambda])$ is a nonlinear matrix-valued interval function in $[\lambda]$. Here, $[\lambda]$ is known as the interval eigenvalue and $[x]$ is the corresponding interval eigenvector. It may be noted that, when $[M]([\lambda])$ is particularly taken as polynomial interval function in $[\lambda]$, the INEP may be referred to as interval polynomial eigenvalue problem. If the degree of the polynomial interval function $[M]([\lambda])$ is k, then the general form of the INEP of degree k may be defined as

$$\begin{aligned} [M]([\lambda])[x] &= \sum_{j=0}^{k} [\lambda]^j [M_j][x] \\ &= \left([M_k][\lambda]^k + [M_{k-1}][\lambda]^{k-1} + \cdots + [M_1][\lambda] + [M_0]\right)[x] = 0, \end{aligned} \tag{7.15}$$

where all the coefficient matrices $[M_j] = [\underline{M_j}, \overline{M_j}]$ for $j = 0, 1, \ldots, k$ are interval square matrices. In this regard, depending upon the degree of the polynomial interval function $[M]([\lambda])$ (value of k), INEPs are classified as follows.

7.4.1 INTERVAL QUADRATIC EIGENVALUE PROBLEM (IQEP)

When the degree of the nonlinear matrix-valued interval function $[M]([\lambda])$ is taken as two (that is $k = 2$), then the INEP may be referred to as an interval quadratic eigenvalue problem (IQEP).

Thus, the IQEP may be expressed as

$$[M]([\lambda])[x] = \left([M_2][\lambda]^2 + [M_1][\lambda] + [M_0]\right)[x] = 0, \tag{7.16}$$

where all the coefficient matrices $[M_2] = \left[\underline{M_2}, \overline{M_2}\right]$, $[M_1] = \left[\underline{M_1}, \overline{M_1}\right]$ and $[M_0] = \left[\underline{M_0}, \overline{M_0}\right]$ are interval square matrices.

7.4.2 INTERVAL CUBIC EIGENVALUE PROBLEM (ICEP)

The case in which the nonlinear matrix-valued interval function $[M]([\lambda])$ in $[\lambda]$ has degree three is known as an interval cubic eigenvalue problem (ICEP) and may be defined as

$$[M]([\lambda])[x] = \left([M_3][\lambda]^3 + [M_2][\lambda]^2 + [M_1][\lambda] + [M_0]\right)[x] = 0, \tag{7.17}$$

where each of the coefficient matrices $\left[M_j\right] = \left[\underline{M_j}, \overline{M_j}\right]$ for $j = 0, 1, 2, 3$ are square matrices, having elements in the form of closed intervals.

Similarly, we may have different types of eigenvalue problems depending upon the degree of the nonlinear interval function.

7.5 SOLUTION OF INEP

The affine arithmetic-based approach has been adopted to compute the interval eigenvalue solutions of the INEP. In this section, the proposed procedure is explained in detail to solve the INEP.

Let us consider an INEP $[M]([\lambda])[x] = 0$ of degree k in which the general form may be written as

$$[M]([\lambda])[x] = \left([M_k][\lambda]^k + [M_{k-1}][\lambda]^{k-1} + \cdots + [M_1][\lambda] + [M_0]\right)[x] = 0. \tag{7.18}$$

Here, all the coefficient matrices $\left[M_j\right] = \left[\underline{M_j}, \overline{M_j}\right]$ for $j = 0, 1, \ldots, k$ are interval matrices of order $l \times l$ having elements $([m_j])_{gh} = \left[\underline{(m_j)_{gh}}, \overline{(m_j)_{gh}}\right]$ for $g, h = 1, 2, \ldots, l$ in the form of closed intervals. Then, the interval coefficient matrices may be expressed as

$$[M_j] = \begin{pmatrix} \left[\underline{(m_j)_{11}}, \overline{(m_j)_{11}}\right] & \left[\underline{(m_j)_{12}}, \overline{(m_j)_{12}}\right] & \cdots & \cdots & \left[\underline{(m_j)_{1l}}, \overline{(m_j)_{1l}}\right] \\ \left[\underline{(m_j)_{21}}, \overline{(m_j)_{21}}\right] & \left[\underline{(m_j)_{22}}, \overline{(m_j)_{22}}\right] & \cdots & \cdots & \left[\underline{(m_j)_{2l}}, \overline{(m_j)_{2l}}\right] \\ \vdots & \vdots & \ddots & & \vdots \\ \vdots & \vdots & & \ddots & \vdots \\ \left[\underline{(m_j)_{l1}}, \overline{(m_j)_{l1}}\right] & \left[\underline{(m_j)_{l2}}, \overline{(m_j)_{l2}}\right] & \cdots & \cdots & \left[\underline{(m_j)_{ll}}, \overline{(m_j)_{ll}}\right] \end{pmatrix},$$

$$\text{for} \quad j = 0, 1, \ldots, k. \tag{7.19}$$

Since the interval coefficient matrices of the kth degree INEP are of order $l \times l$, then there exists kl number of interval eigenvalues of the INEP given in Eq. (7.18). To find these eigenvalues, we have adopted an affine arithmetic approach. In this regard, all the interval coefficient (Eq. (7.19)) matrices have been transformed into affine form representations (as given in Chapter 3). After the conversion, the INEP is transformed into affine nonlinear eigenvalue problem (ANEP), whose parameters can be treated as crisp (or exact) numbers. Therefore, the ANEP may be linearized and further solved as described below.

The respective center and half-width (or radius) matrices of the above interval coefficient matrices can be found as below:

$$(M_j)_c = \begin{pmatrix} (m_j)_{11}{}^{(0)} & (m_j)_{12}{}^{(0)} & \cdots & \cdots & (m_j)_{1l}{}^{(0)} \\ (m_j)_{21}{}^{(0)} & (m_j)_{22}{}^{(0)} & \cdots & \cdots & (m_j)_{2l}{}^{(0)} \\ \vdots & \vdots & \ddots & & \vdots \\ \vdots & \vdots & & \ddots & \vdots \\ (m_j)_{l1}{}^{(0)} & (m_j)_{l2}{}^{(0)} & \cdots & \cdots & (m_j)_{ll}{}^{(0)} \end{pmatrix},$$

$$(M_j)_\Delta = \begin{pmatrix} (m_j)_{11}{}^{(1)} & (m_j)_{12}{}^{(1)} & \cdots & \cdots & (m_j)_{1l}{}^{(1)} \\ (m_j)_{21}{}^{(1)} & (m_j)_{22}{}^{(1)} & \cdots & \cdots & (m_j)_{2l}{}^{(1)} \\ \vdots & \vdots & \ddots & & \vdots \\ \vdots & \vdots & & \ddots & \vdots \\ (m_j)_{l1}{}^{(1)} & (m_j)_{l2}{}^{(1)} & \cdots & \cdots & (m_j)_{ll}{}^{(1)} \end{pmatrix}, \tag{7.20}$$

where $(m_j)_{gh}{}^{(0)} = \frac{1}{2}\left(\underline{(m_j)_{gh}} + \overline{(m_j)_{gh}}\right)$ and $(m_j)_{gh}{}^{(1)} = \frac{1}{2}\left(\overline{(m_j)_{gh}} - \underline{(m_j)_{gh}}\right)$ for $j = 0, 1, \ldots, k$ and $g, h = 1, \ldots, l$.

Thus, the affine form representations of the interval coefficient matrices ((7.19)) by using Eq. (7.20) are determined as

$$\hat{M}_j(\varepsilon_{j,gh}) = \begin{pmatrix} (m_j)_{11}{}^{(0)} + (m_j)_{11}{}^{(1)}\varepsilon_{j,11} & (m_j)_{12}{}^{(0)} + (m_j)_{12}{}^{(1)}\varepsilon_{j,12} & \cdots & \cdots & (m_j)_{1l}{}^{(0)} + (m_j)_{1l}{}^{(1)}\varepsilon_{j,1l} \\ (m_j)_{21}{}^{(0)} + (m_j)_{21}{}^{(1)}\varepsilon_{j,21} & (m_j)_{22}{}^{(0)} + (m_j)_{22}{}^{(1)}\varepsilon_{j,22} & \cdots & \cdots & (m_j)_{2l}{}^{(0)} + (m_j)_{2l}{}^{(1)}\varepsilon_{j,2l} \\ \vdots & \vdots & \ddots & & \vdots \\ \vdots & \vdots & & \ddots & \vdots \\ (m_j)_{l1}{}^{(0)} + (m_j)_{l1}{}^{(1)}\varepsilon_{j,l1} & (m_j)_{l2}{}^{(0)} + (m_j)_{l2}{}^{(1)}\varepsilon_{j,l2} & \cdots & \cdots & (m_j)_{ll}{}^{(0)} + (m_j)_{ll}{}^{(1)}\varepsilon_{j,ll} \end{pmatrix}, \tag{7.21}$$

where $\varepsilon_{j,gh} \in [-1, 1]$ for $j = 1, \ldots, k$ and $g, h = 1, \ldots, l$ are noise symbols of each element of all the coefficient matrices.

Hence, the INEP is converted into its affine form representation and may be called an affine nonlinear eigenvalue problem (ANEP). Then, the ANEP may be expressed as

$$\begin{aligned} \hat{M}(\varepsilon^*)(\hat{\lambda}(\varepsilon^*))\hat{x}(\varepsilon^*) &= \sum_{j=0}^{k}\{\hat{\lambda}^j(\varepsilon^*)\hat{M}_j(\varepsilon_{j,gh})\}\hat{x}(\varepsilon^*) = 0 \\ &\Rightarrow \{\hat{M}_0(\varepsilon_{0,gh}) + \hat{M}_1(\varepsilon_{1,gh})\tilde{\lambda}(\varepsilon^*) + \cdots + \hat{M}_{k-1}(\varepsilon_{k-1,gh})\hat{\lambda}^{k-1}(\varepsilon^*) \\ &\quad + \hat{M}_k(\varepsilon_{k,gh})\hat{\lambda}^k(\varepsilon^*)\}\hat{x}(\varepsilon^*) = 0, \end{aligned} \tag{7.22}$$

where ε^* may be either a newly generated noise symbol or a function of existing noise symbols $\varepsilon_{j,gh} \in [-1, 1]$ for $j = 0, 1, \ldots, k$ and $g, h = 1, \ldots, l$. Moreover, $\hat{\lambda}(\varepsilon^*)$ and $\hat{x}(\varepsilon^*)$ are the corresponding affine forms of interval eigenvalue $[\lambda]$ and interval eigenvector $[x]$ of the INEP (7.18), respectively.

Here, all the parameters of the ANEP are in the form of affine representations. Thus, ANEP is linearized by adopting the procedure given in Section 7.2 and is transformed into an affine generalized eigenvalue problem (AGEP) as

$$\hat{G}\left(\varepsilon^*\right)\hat{x}^*\left(\varepsilon^*\right) = \hat{\lambda}\left(\varepsilon^*\right)\hat{H}\left(\varepsilon^*\right)\hat{x}^*\left(\varepsilon^*\right), \tag{7.23}$$

where the coefficient affine matrices of the AGEP are given as follows:

$$\hat{G}\left(\varepsilon^{*}\right)=\left(\begin{array}{c|ccc} -\hat{M}_{0}\left(\varepsilon_{0,gh}\right) & 0 & \cdots & 0 \\ \hline 0 & I & & 0 \\ \vdots & \vdots & \ddots & \vdots \\ 0 & 0 & \cdots & I \end{array}\right)_{kl\times kl}$$

and

$$\hat{H}\left(\varepsilon^{*}\right)=\left(\begin{array}{cccc|c} \hat{M}_{1}\left(\varepsilon_{1,gh}\right) & \cdots & \hat{M}_{k-1}(\varepsilon_{k-1,gh}) & & N_{k}(\varepsilon_{k,gh}) \\ \hline I & & 0 & & 0 \\ \vdots & \ddots & \vdots & & \vdots \\ 0 & \cdots & I & & 0 \end{array}\right)_{kl\times kl}. \tag{7.24}$$

After the linearization of ANEP the problem (7.22) into AGEP (7.23), the affine eigenvalue solutions in terms of noise symbols are obtained by solving

$$\det\left(\hat{G}(\varepsilon^{*})-\hat{\lambda}\hat{H}(\varepsilon^{*})\right)=0 \tag{7.25}$$

symbolically. Thus, all the obtained solutions are in symbolic form with symbols $\varepsilon_{j,gh}\in[-1,1]$ for $j=0,1,\ldots,k$ and $g,h=1,\ldots,l$, which is the required eigenvalue solutions in affine forms $\hat{\lambda}_i(\varepsilon^{*})$ for $i=1,2,\ldots,kl$.

Now, because every noise symbol varies from -1 to 1, the interval bounds of the eigenvalues $[\lambda_i]=[\underline{\lambda_i},\overline{\lambda_i}]$ may be calculated as follows:

- Lower bounds:

$$\underline{\lambda_i}=\min_{\varepsilon^{*}\in[-1,1]}\hat{\lambda}_i\left(\varepsilon^{*}\right); \tag{7.26}$$

- Upper bounds:

$$\overline{\lambda_i}=\max_{\varepsilon^{*}\in[-1,1]}\hat{\lambda}_i\left(\varepsilon^{*}\right); \tag{7.27}$$

for $i=1,2,\ldots,kl$.

Example 7.5 Determine the lower and upper eigenvalue bounds of the IQEP $([\lambda]^2[M_2]+[\lambda][M_1]+[M_0])[x]=0$, where all the coefficient matrices of the problem may be taken as 3×3

interval matrices given as follows:

$$[M_2] = \begin{pmatrix} [1.8, 2.2] & 0 & 0 \\ 0 & [1.8, 2.2] & 0 \\ 0 & 0 & [1.8, 2.2] \end{pmatrix},$$

$$[M_1] = \begin{pmatrix} [11.7, 12.3] & [-4.1, -3.9] & 0 \\ [-4.1, -3.9] & [11.7, 12.3] & [-4.1, -3.9] \\ 0 & [-4.1, -3.9] & [11.7, 12.3] \end{pmatrix}$$

and

$$[M_0] = \begin{pmatrix} [8.7, 9.3] & [-3.1, -2.9] & 0 \\ [-3.1, -2.9] & [8.7, 9.3] & [-3.1, -2.9] \\ 0 & [-3.1, -2.9] & [8.7, 9.3] \end{pmatrix}.$$

Solution: In this problem, the degree of the INEP is two (that is, $k = 2$) and the dimension of all the coefficient matrices is three (that is $l = 3$). Thus, the given IQEP has $kl = 2 \times 3 = 6$ number of eigenvalues. First of all, we have to convert the given interval coefficient matrices into their affine form representation. Thus, after converting each element of the given matrix into affine form (as given in Eq. (7.21) of Section 7.5), we may have

$$\hat{M}_2\,(\varepsilon_i)_{i=1,2,3} = \begin{pmatrix} 2 + 0.2\varepsilon_1 & 0 & 0 \\ 0 & 2 + 0.2\varepsilon_2 & 0 \\ 0 & 0 & 2 + 0.2\varepsilon_3 \end{pmatrix},$$

$$\hat{M}_1(\varepsilon_i)_{i=4,\ldots 10} = \begin{pmatrix} 12 + 0.3\varepsilon_4 & -4 + 0.1\varepsilon_5 & 0 \\ -4 + 0.1\varepsilon_6 & 12 + 0.3\varepsilon_7 & -4 + 0.1\varepsilon_8 \\ 0 & -4 + 0.1\varepsilon_9 & 12 + 0.3\varepsilon_{10} \end{pmatrix}$$

and

$$\hat{M}_0(\varepsilon_i)_{i=11,\ldots,17} = \begin{pmatrix} 9 + 0.3\varepsilon_{11} & -3 + 0.1\varepsilon_{12} & 0 \\ -3 + 0.1\varepsilon_{13} & 9 + 0.3\varepsilon_{14} & -3 + 0.1\varepsilon_{15} \\ 0 & -3 + 0.1\varepsilon_{16} & 9 + 0.3\varepsilon_{17} \end{pmatrix}.$$

Now, utilizing the linearization procedure given in Section 7.2, the ANEP is linearized into AGEP $\hat{G}\hat{x}^* = \hat{\lambda}\hat{H}\hat{x}^*$, where

$$\hat{G} = \left(\begin{array}{c|c} -\hat{M}_0(\varepsilon_i)_{i=11,\ldots,17} & 0 \\ \hline 0 & I \end{array}\right) \text{ and } \hat{H} = \left(\begin{array}{c|c} \hat{M}_1(\varepsilon_i)_{i=4,\ldots,10} & \hat{M}_2(\varepsilon_i)_{i=1,2,3} \\ \hline I & 0 \end{array}\right).$$

Table 7.3: Eigenvalue bounds ($[\lambda]$) of the IQEP for Example 7.5

i	Interval Eigenvalue Bounds			Crisp Eigenvalues (λ)
	Center (λ_c)	Lower ($\underline{\lambda_i}$)	Upper ($\overline{\lambda_i}$)	
1	-8.0008	-8.9050	-7.2587	-8.0008
2	-5.1213	-5.6436	-4.6895	-5.1213
3	-1.9546	-2.1777	-1.6624	-1.9546
4	-1.2169	-1.4214	-1.1010	-1.2169
5	-0.8787	-0.9014	-0.8564	-0.8787
6	-0.8276	-0.8392	-0.8163	-0.8276

Finally, the interval eigenvalue solutions of the given IQEP are evaluated by adopting the proposed method given in Section 7.5. All the interval bounds of the eigenvalue solutions are listed in Table 7.3.

It may be clearly observed from the above table that, the crisp eigenvalue solutions lie in between the lower- and upper-bound solutions of the IQEP and coincide with the central value of the interval solutions.

7.6 FUZZY NONLINEAR EIGENVALUE PROBLEM (FNEP)

When uncertain parameters are considered in the form of fuzzy numbers, the NEP may be referred to as FNEP and is denoted as follows:

$$\tilde{M}\left(\tilde{\lambda}\right)\tilde{x} = 0, \tag{7.28}$$

where $\tilde{\lambda}$ is the fuzzy eigenvalue, $\tilde{x}$ is the corresponding fuzzy eigenvector, and $\tilde{M}(\tilde{\lambda})$ is the fuzzy matrix-valued nonlinear function in $\tilde{\lambda}$. Then, the general form of the FNEP of degree "k" may be considered in the following form.

$$\tilde{M}\left(\tilde{\lambda}\right)\tilde{x} = \sum_{j=0}^{k} \tilde{\lambda}^j \tilde{M}_j \tilde{x} = \left(\tilde{M}_k \tilde{\lambda}^k + \tilde{M}_{k-1}\tilde{\lambda}^{k-1} + \cdots + \tilde{M}_1\tilde{\lambda} + \tilde{M}_0\right)\tilde{x} = 0, \tag{7.29}$$

where each of the coefficients $\tilde{N}_j$ for $j = 0, 1, \ldots, k$ are fuzzy square matrices (the elements may be taken in the form of TFNs or TrFNs). In a similar manner, as given for crisp and interval parameters, FNEP may also be classified (depending upon the degree of the nonlinear matrix-valued fuzzy function $\tilde{M}(\tilde{\lambda})$) in the following manner.

7.6.1 FUZZY QUADRATIC EIGENVALUE PROBLEM (FQEP)

When the degree of the nonlinear matrix-valued fuzzy function $\tilde{M}(\tilde{\lambda})$ is taken as two (that is $k = 2$), then the FNEP may be referred as a fuzzy quadratic eigenvalue problem (FQEP). Thus, the FQEP may be expressed as

$$\tilde{M}\left(\tilde{\lambda}\right)\tilde{x} = \left(\tilde{M}_2\tilde{\lambda}^2 + \tilde{M}_1\tilde{\lambda} + \tilde{M}_0\right)\tilde{x} = 0, \tag{7.30}$$

where all the coefficient matrices $\tilde{M}_2$, $\tilde{M}_1$, and $\tilde{M}_0$ are fuzzy square matrices.

7.6.2 FUZZY CUBIC EIGENVALUE PROBLEM (FCEP)

The case in which the nonlinear matrix-valued interval function $\tilde{M}\left(\tilde{\lambda}\right)$ in $\tilde{\lambda}$ has degree three is known as fuzzy cubic eigenvalue problem (FCEP) and may be defined as

$$\tilde{M}\left(\tilde{\lambda}\right)\tilde{x} = \left(\tilde{M}_3\tilde{\lambda}^3 + \tilde{M}_2\tilde{\lambda}^2 + \tilde{M}_1\tilde{\lambda} + \tilde{M}_0\right)\tilde{x} = 0, \tag{7.31}$$

where each of the coefficient matrices $\tilde{M}_j$ for $j = 0, 1, 2, 3$ are square matrices having elements in the form of fuzzy numbers viz. TFN and TrFN, and so on.

7.7 SOLUTION OF FNEP

In this section, a fuzzy-affine approach is proposed to find the fuzzy solutions of the FNEP efficiently. In this regard, let us consider the general form of a kth degree FNEP (from Eq. (7.29)) given as follows:

$$\tilde{M}\left(\tilde{\lambda}\right)\tilde{x} = \sum_{j=0}^{k}\tilde{\lambda}^j\tilde{M}_j\tilde{x} = \left(\tilde{M}_k\tilde{\lambda}^k + \tilde{M}_{k-1}\tilde{\lambda}^{k-1} + \cdots + \tilde{M}_1\tilde{\lambda} + \tilde{M}_0\right)\tilde{x} = 0, \tag{7.32}$$

where all the coefficient matrices $\tilde{M}_j$ for $j = 0, 1, \ldots, k$ are fuzzy square matrices of order $l \times l$ having elements $(\tilde{m}_j)_{gh}$ for $g, h = 1, 2, \ldots, l$ in the form of fuzzy numbers (particularly, in the form of TFNs or TrFNs). Then, the fuzzy coefficient matrices may be written as

$$\tilde{M}_j = \begin{pmatrix} (\tilde{m}_j)_{11} & (\tilde{m}_j)_{12} & \cdots & \cdots & (\tilde{m}_j)_{1l} \\ (\tilde{m}_j)_{21} & (\tilde{m}_j)_{22} & \cdots & \cdots & (\tilde{m}_j)_{2l} \\ \vdots & \vdots & \ddots & & \vdots \\ \vdots & \vdots & & \ddots & \vdots \\ (\tilde{m}_j)_{l1} & (\tilde{m}_j)_{l2} & \cdots & \cdots & (\tilde{m}_j)_{ll} \end{pmatrix}, \quad \text{for} \quad j = 0, 1, \ldots, k. \tag{7.33}$$

Adopting the parameterization of each fuzzy number present in all the coefficient matrices as given in Chapter 4 (by using α-cut approach), the FNEP Eq. (7.32) may be converted into fuzzy parametric NEP given as

$$\tilde{M}\left(\tilde{\lambda}(a)\right)\tilde{x}(a) = \sum_{j=0}^{k}\left\{\tilde{\lambda}^{j}(a)\tilde{M}_j(a)\right\}\tilde{x}(a)$$
$$= \left\{\tilde{M}_0(a) + \tilde{M}_1(a)\tilde{\lambda}(a) + \cdots + \tilde{M}_k(a)\tilde{\lambda}^k(a)\right\}\tilde{x}(a) = 0, \tag{7.34}$$

where the fuzzy parametric coefficient matrices is expressed as follows:

$$\tilde{M}_j(a) = \begin{pmatrix} \left[\underline{(m_j(a))_{11}}, \overline{(m_j(a))_{11}}\right] & \left[\underline{(m_j(a))_{12}}, \overline{(m_j(a))_{12}}\right] & \cdots & \cdots & \left[\underline{(m_j(a))_{1l}}, \overline{(m_j(a))_{1l}}\right] \\ \left[\underline{(m_j(a))_{21}}, \overline{(m_j(a))_{21}}\right] & \left[\underline{(m_j(a))_{22}}, \overline{(m_j(a))_{22}}\right] & \cdots & \cdots & \left[\underline{(m_j(a))_{2l}}, \overline{(m_j(a))_{2l}}\right] \\ \vdots & \vdots & \ddots & & \vdots \\ \vdots & \vdots & & \ddots & \vdots \\ \left[\underline{(m_j(a))_{l1}}, \overline{(m_j(a))_{l1}}\right] & \left[\underline{(m_j(a))_{l2}}, \overline{(m_j(a))_{l2}}\right] & \cdots & \cdots & \left[\underline{(m_j(a))_{ll}}, \overline{(m_j(a))_{ll}}\right] \end{pmatrix},$$

$$\text{for} \quad a \in [0, 1]. \tag{7.35}$$

Moreover, the fuzzy parametric NEP is again converted into its affine form representation as given in Chapters 3 and 4. Thus, the fuzzy-affine NEP of degree k may be written as

$$\hat{M}\left(a, \varepsilon^*\right)\left(\hat{\lambda}\left(a, \varepsilon^*\right)\right)\hat{x}\left(a, \varepsilon^*\right) = \sum_{j=0}^{k}\left\{\hat{\lambda}^j\left(a, \varepsilon^*\right)\hat{M}_j\left(a, \varepsilon_{j,pq}\right)\right\}\hat{x}\left(a, \varepsilon^*\right) = 0$$
$$\Rightarrow \left\{\hat{M}_0\left(a, \varepsilon_{0,gh}\right) + \hat{M}_1\left(a, \varepsilon_{1,gh}\right)\hat{\lambda}\left(a, \varepsilon^*\right) + \cdots\right.$$
$$\left.\cdots + \hat{M}_k\left(a, \varepsilon_{k,gh}\right)\hat{\lambda}^k\left(a, \varepsilon^*\right)\right\}\hat{x}\left(a, \varepsilon^*\right) = 0, \tag{7.36}$$

where ε^* may be either a newly generated noise symbol or a function of existing noise symbols $\varepsilon_{j,gh}$. Here, all the coefficient matrices are in the form of fuzzy-affine representations and may be written as

$$\hat{M}_j\left(a, \varepsilon_{j,gh}\right) = \begin{pmatrix} \left(\hat{m}_j\left(a, \varepsilon_{j,11}\right)\right)_{11} & \left(\hat{m}_j\left(a, \varepsilon_{j,12}\right)\right)_{12} & \cdots & \cdots & \left(\hat{m}_j\left(a, \varepsilon_{j,1l}\right)\right)_{1l} \\ \left(\hat{m}_j\left(a, \varepsilon_{j,21}\right)\right)_{21} & \left(\hat{m}_j\left(a, \varepsilon_{j,22}\right)\right)_{22} & \cdots & \cdots & \left(\hat{m}_j\left(a, \varepsilon_{j,2l}\right)\right)_{2l} \\ \vdots & \vdots & \ddots & & \vdots \\ \vdots & \vdots & & \ddots & \vdots \\ \left(\hat{m}_j\left(a, \varepsilon_{j,l1}\right)\right)_{l1} & \left(\hat{m}_j\left(a, \varepsilon_{j,l2}\right)\right)_{l2} & \cdots & \cdots & \left(\hat{m}_j\left(a, \varepsilon_{j,ll}\right)\right)_{ll} \end{pmatrix}, \quad (7.37)$$

where $a \in [0,1]$ and $\varepsilon_{j,gh} \in [-1,1]$ for $j = 0,1,\ldots,k$ and $g,h = 1,\ldots,l$. $\varepsilon_{j,gh}$ for $j = 0,1,\ldots,k$ and $g,h = 1,\ldots,l$ are different noise symbols for different affine form representations of the elements of all the coefficient matrices. Further,

$$\left(\hat{m}_j\left(a, \varepsilon_{j,gh}\right)\right)_{gh} = \left(m_j\right)_{gh}{}^{(0)} + \left(m_j\right)_{gh}{}^{(1)} \varepsilon_{j,gh}, \quad (7.38)$$

where $\left(m_j\right)_{gh}{}^{(0)} = \frac{1}{2}\left(\underline{(m_j(a))_{gh}} + \overline{(m_j(a))_{gh}}\right)$ and $\left(m_j\right)_{gh}{}^{(1)} = \frac{1}{2}\left(\overline{(m_j(a))_{gh}} - \underline{(m_j(a))_{gh}}\right)$, for $j = 0,1,\ldots,k$ and $g,h = 1,\ldots,l$.

Moreover, $\hat{\lambda}(a,\varepsilon^*)$ and $\hat{x}(a,\varepsilon^*)$ are the corresponding fuzzy-affine forms of fuzzy eigenvalue $\tilde{\lambda}$ and fuzzy eigenvector $\tilde{x}$ of the FNEP, respectively.

Thus, the FNEP is transformed into a form involving several parameters such as the fuzzy parameter (a) and the noise symbols (ε^*). Hence, it can be linearized into a GEP.

Utilizing the linearization procedure for the NEP given in Section 7.2, the fuzzy-affine NEP may be transformed into a fuzzy-affine GEP as

$$\hat{G}\left(a,\varepsilon^*\right)\hat{x}^*\left(a,\varepsilon^*\right) = \hat{\lambda}\left(a,\varepsilon^*\right)\hat{H}\left(a,\varepsilon^*\right)\hat{x}^*\left(a,\varepsilon^*\right), \quad (7.39)$$

where

$$\hat{G}\left(a,\varepsilon^*\right) = \left(\begin{array}{c|ccc} -\hat{M}_0\left(a,\varepsilon_{0,gh}\right) & 0 & \cdots & 0 \\ \hline 0 & I & & 0 \\ \vdots & \vdots & \ddots & \vdots \\ 0 & 0 & \cdots & I \end{array}\right)_{kl\times kl}$$

and

$$\hat{H}\left(a,\varepsilon^*\right) = \left(\begin{array}{ccc|c} \hat{M}_1\left(a,\varepsilon_{1,gh}\right) & \cdots & \hat{M}_{k-1}\left(a,\varepsilon_{k-1,gh}\right) & \hat{M}_k\left(a,\varepsilon_{k,gh}\right) \\ \hline I & & 0 & 0 \\ \vdots & \ddots & \vdots & \vdots \\ 0 & \cdots & I & 0 \end{array}\right)_{kl\times kl}. \quad (7.40)$$

The above fuzzy-affine GEP (7.39) is solved and the required eigenvalue solutions in fuzzy-affine forms $\hat{\lambda}_i(a, \varepsilon^*)$ for $i = 1, 2, \ldots, kl$ may be computed.

Since every consisting noise symbols ($\varepsilon_{j,gh}$ for $j = 0, 1, \ldots, k$ and $g, h = 1, \ldots, l$) vary from -1 to 1, the fuzzy parametric eigenvalue solutions $\tilde{\lambda}_i(a) = [\underline{\lambda_i(a)}, \overline{\lambda_i(a)}]$ are calculated as follows:

- Lower bounds:

$$\underline{\lambda_i(a)} = \min_{\varepsilon^* \in [-1,1]} \hat{\lambda}_i\left(a, \varepsilon^*\right); \tag{7.41}$$

- Upper bounds:

$$\overline{\lambda_i(a)} = \max_{\varepsilon^* \in [-1,1]} \hat{\lambda}_i\left(a, \varepsilon^*\right); \tag{7.42}$$

for $i = 1, 2, \ldots, kl$.

Finally, all the fuzzy solutions of the FNEP of degree k can be evaluated by varying its fuzzy parameter (a) from 0 to 1 and the fuzzy solution plots can also be constructed by putting different values of a in the lower and upper bounds of the fuzzy parametric solutions given in Eqs. (7.41) and (7.42).

Example 7.6 Find the solution of an FQEP $\left(\tilde{\lambda}^2 \tilde{M}_2 + \tilde{\lambda}\tilde{M}_1 + \tilde{M}_0\right)\tilde{x} = 0$, where its coefficients are 3×3 fuzzy matrices having elements in the form of TFNs given as below. Also, plot all the fuzzy eigenvalue solutions of the given problem.

$$\tilde{M}_2 = \begin{bmatrix} (1.8, 2, 2.2) & 0 & 0 \\ 0 & (1.8, 2, 2.2) & 0 \\ 0 & 0 & (1.8, 2, 2.2) \end{bmatrix};$$

$$\tilde{M}_1 = \begin{bmatrix} (11.7, 12, 12.3) & (-4.1, -4, -3.9) & 0 \\ (-4.1, -4, -3.9) & (11.7, 12, 12.3) & (-4.1, -4, -3.9) \\ 0 & (-4.1, -4, -3.9) & (11.7, 12, 12.3) \end{bmatrix};$$

$$\tilde{M}_0 = \begin{bmatrix} (8.7, 9, 9.3) & (-3.1, -3, -2.9) & 0 \\ (-3.1, -3, -2.9) & (8.7, 9, 9.3) & (-3.1, -3, -2.9) \\ 0 & (-3.1, -3, -2.9) & (8.7, 9, 9.3) \end{bmatrix}.$$

Solution: Here, the constituting elements of the given coefficient matrices are in the form of TFNs. Then, parameterizing the given fuzzy matrices $\tilde{M}_2$, $\tilde{M}_1$, and $\tilde{M}_0$ by using a-cut approach, we may have

$$\tilde{M}_2(a) = \begin{pmatrix} [1.8+0.2a, 2.2-0.2a] & 0 & 0 \\ 0 & [1.8+0.2a, 2.2-0.2a] & 0 \\ 0 & 0 & [1.8+0.2a, 2.2-0.2a] \end{pmatrix};$$

$$\tilde{M}_1(a) = \begin{pmatrix} [11.7+0.3a, 12.3-0.3a] & [-4.1+0.1a, -3.9-0.1a] & 0 \\ [-4.1+0.1a, -3.9-0.1a] & [11.7+0.3a, 12.3-0.3a] & [-4.1+0.1a, -3.9-0.1a] \\ 0 & [-4.1+0.1a, -3.9-0.1a] & [11.7+0.3a, 12.3-0.3a] \end{pmatrix};$$

$$\tilde{M}_0(a) = \begin{pmatrix} [8.7+0.3a, 9.3-0.3a] & [-3.1+0.1a, -2.9-0.1a] & 0 \\ [-3.1+0.1a, -2.9-0.1a] & [8.7+0.3a, 9.3-0.3a] & [-3.1+0.1a, -2.9-0.1a] \\ 0 & [-3.1+0.1a, -2.9-0.1a] & [8.7+0.3a, 9.3-0.3a] \end{pmatrix}.$$

Further, the parametric fuzzy numbers are changed to their fuzzy-affine forms by adopting the procedure given in Section 7.7. Thus, the FNEP with different parameters and the fuzzy-affine coefficients are in the form

$$\hat{M}_2\,(a, \varepsilon_i)_{i=1,2,3} = \begin{pmatrix} 2+(0.2-0.2a)\varepsilon_1 & 0 & 0 \\ 0 & 2+(0.2-0.2a)\varepsilon_2 & 0 \\ 0 & 0 & 2+(0.2-0.2a)\varepsilon_3 \end{pmatrix};$$

$$\hat{M}_1\,(a, \varepsilon_i)_{i=4,\ldots,10} = \begin{pmatrix} 12+(0.3-0.3a)\varepsilon_4 & -4+(0.1-0.1a)\varepsilon_5 & 0 \\ -4+(0.1-0.1a)\varepsilon_6 & 12+(0.3-0.3a)\varepsilon_7 & -4+(0.1-0.1a)\varepsilon_8 \\ 0 & -4+(0.1-0.1a)\varepsilon_9 & 12+(0.3-0.3a)\varepsilon_{10} \end{pmatrix};$$

$$\hat{M}_0\,(a, \varepsilon_i)_{i=11,\ldots,17} = \begin{pmatrix} 9+(0.3-0.3a)\varepsilon_{11} & -3+(0.1-0.1a)\varepsilon_{12} & 0 \\ -3+(0.1-0.1a)\varepsilon_{13} & 9+(0.3-0.3a)\varepsilon_{14} & -3+(0.1-0.1a)\varepsilon_{15} \\ 0 & -3+(0.1-0.1a)\varepsilon_{16} & 9+(0.3-0.3a)\varepsilon_{17} \end{pmatrix}.$$

Table 7.4: Triangular fuzzy eigenvalue bounds ($\tilde{\lambda}$) of Example 7.6 for different values of a.

a	Lower and Upper Bounds	λ_1	λ_2	λ_3	λ_4	λ_5	λ_6
0	$\underline{\lambda}$	-8.9050	-5.6436	-2.1777	-1.4214	-0.9014	-0.8392
	$\overline{\lambda}$	-7.2587	-4.6895	-1.6624	-1.1010	-0.8564	-0.8163
0.2	$\underline{\lambda}$	-8.7087	-5.5305	-2.1336	-1.3584	-0.8968	-0.8368
	$\overline{\lambda}$	-7.3964	-4.7698	-1.7417	-1.1219	-0.8608	-0.8185
0.4	$\underline{\lambda}$	-8.5206	-5.4220	-2.0895	-1.3137	-0.8923	-0.8345
	$\overline{\lambda}$	-7.5391	-4.8530	-1.8033	-1.1436	-0.8653	-0.8208
0.7	$\underline{\lambda}$	-8.2528	-5.2672	-2.0229	-1.2608	-0.8854	-0.8310
	$\overline{\lambda}$	-7.7633	-4.9835	-1.8827	-1.1785	-0.8720	-0.8241
0.9	$\underline{\lambda}$	-8.0832	-5.1690	-1.9777	-1.2308	-0.8809	-0.8287
	$\overline{\lambda}$	-7.9201	-5.0745	-1.9312	-1.2036	-0.8764	-0.8264
1	$\underline{\lambda}$	-8.0008	-5.1213	-1.9546	-1.2169	-0.8787	-0.8276
$(\underline{\lambda} = \overline{\lambda})$	$\overline{\lambda}$	-8.0008	-5.1213	-1.9546	-1.2169	-0.8787	-0.8276

Now, utilizing the linearization procedure given in Section 7.2, the fuzzy-affine NEP is linearized into fuzzy-affine GEP $\hat{G}\hat{x}^* = \hat{\lambda}\hat{H}\hat{x}^*$, where

$$\hat{G} = \begin{pmatrix} -\hat{M}_0(a, \varepsilon_i)_{i=11,\dots,17} & 0 \\ 0 & I \end{pmatrix} \text{ and } \hat{H} = \begin{pmatrix} \hat{M}_1(a, \varepsilon_i)_{i=4,\dots,10} & \hat{M}_2(a, \varepsilon_i)_{i=1,2,3} \\ I & 0 \end{pmatrix}.$$

Then the fuzzy eigenvalues in the form of TFN are evaluated by adopting the proposed method given in Section 7.7. All the fuzzy eigenvalue plots are depicted in Figures 7.2–7.7.

Lastly, for some values of the fuzzy parameter viz. $a = 0, 0.2, 0.4, 0.7, 0.9$, and 1, the eigenvalues of Example 7.6 are listed in Table 7.4.

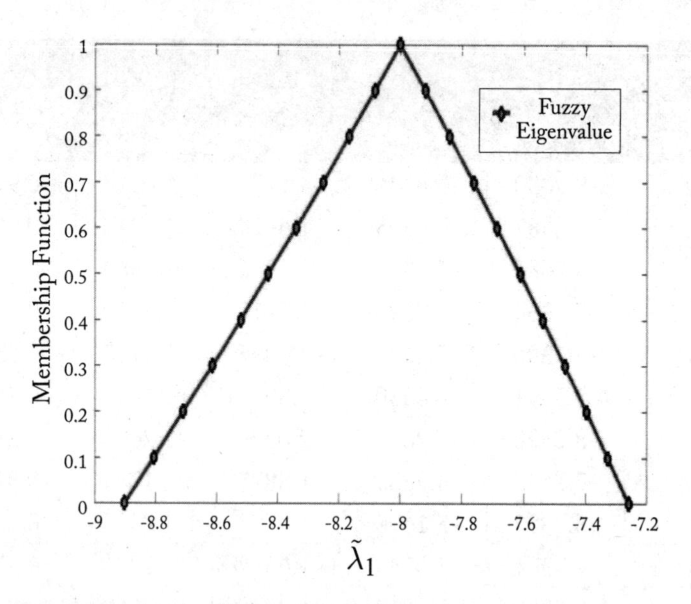

Figure 7.2: First eigenvalue plot for Example 7.6.

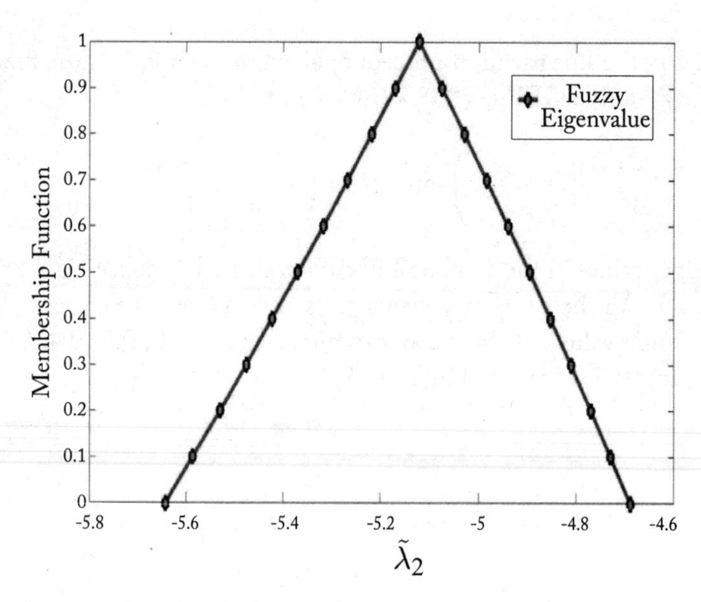

Figure 7.3: Second eigenvalue plot for Example 7.6.

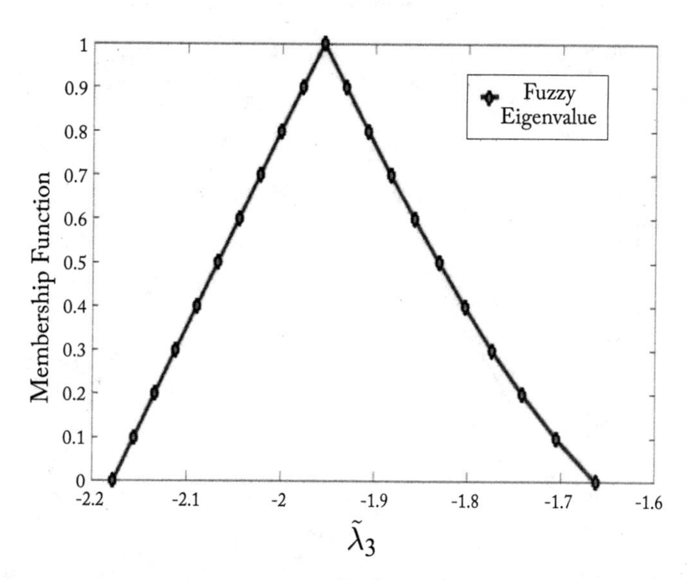

Figure 7.4: Third eigenvalue plot for Example 7.6.

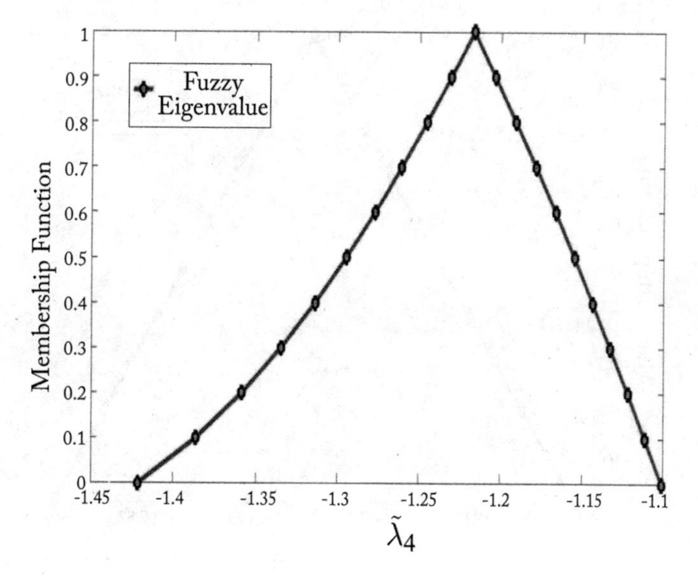

Figure 7.5: Fourth eigenvalue plot for Example 7.6.

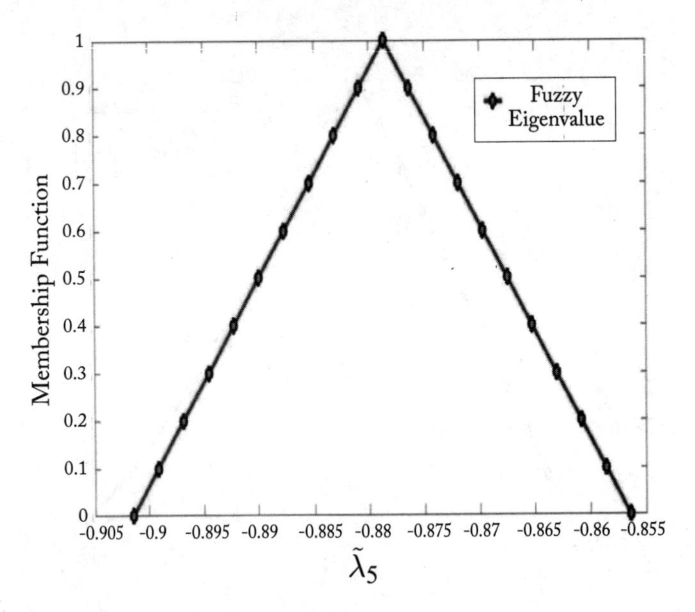

Figure 7.6: Fifth eigenvalue plot for Example 7.6.

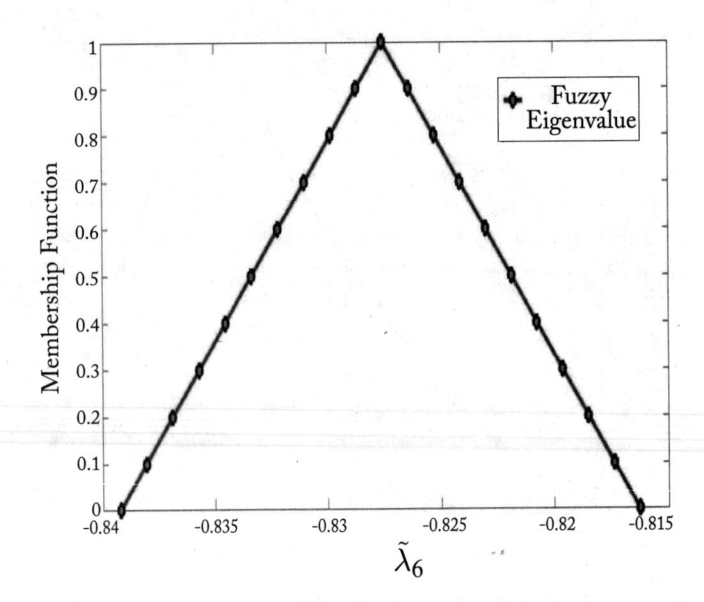

Figure 7.7: Sixth eigenvalue plot for Example 7.6.

7.8 EXERCISES

7.1. Compute all the eigenvalue solutions of the crisp quadratic eigenvalue problem $(A\lambda^2 + B\lambda + C)x = 0$, where

$$A = \begin{pmatrix} 2 & 0 & 0 \\ 0 & 2 & 0 \\ 0 & 0 & 2 \end{pmatrix}, \quad B = \begin{pmatrix} 12 & -4 & 0 \\ -4 & 12 & -4 \\ 0 & -4 & 12 \end{pmatrix}, \quad \text{and } C = \begin{pmatrix} 9 & -3 & 0 \\ -3 & 9 & -3 \\ 0 & -3 & 9 \end{pmatrix}.$$

7.2. Evaluate the lower and upper bounds of the interval eigenvalue solutions for the following interval quadratic eigenvalue problem (crisp form given in Example 7.4).

$$\left([A][\lambda]^2 + [B][\lambda] + [C]\right)[x] = 0,$$

where

$$[A] = \begin{pmatrix} 1 & [0.07, 0.27] & [-0.35, -0.15] & [0.50, 0.58] \\ [0.44, 0.50] & 1 & [0.65, 0.69] & [-0.34, -0.30] \\ [-0.15, -0.07] & [0.30, 0.40] & 1 & [-0.78, -0.70] \\ [0.50, 0.60] & [0.40, 0.46] & [0.30, 0.42] & 1 \end{pmatrix},$$

$$[B] = \begin{pmatrix} [0.20, 0.24] & [0.01, 0.03] & [0.10, 0.14] & [0.13, 0.15] \\ [0.01, 0.03] & [0.12, 0.16] & [0.03, 0.05] & [-0.08, -0.04] \\ [0.10, 0.14] & [0.03, 0.05] & [0.24, 0.32] & [0.04, 0.12] \\ [0.13, 0.15] & [-0.08, -0.04] & [0.04, 0.12] & [0.22, 0.30] \end{pmatrix}, \text{ and}$$

$$[C] = \begin{pmatrix} -3.0475 & -2.1879 & -1.9449 & -2.8242 \\ -2.6500 & -2.4724 & -2.3515 & -2.1053 \\ -0.7456 & -0.6423 & -1.3117 & -0.1852 \\ -4.0500 & -3.0631 & -2.8121 & -3.7794 \end{pmatrix}.$$

Also, verify whether the interval solutions contain the crisp solutions given in Example 7.4 or not.

7.3. Determine the trapezoidal fuzzy solution plots of the fuzzy quadratic eigenvalue problem $(\tilde{A}\tilde{\lambda}^2 + \tilde{B}\tilde{\lambda} + \tilde{C})\tilde{x} = 0$, where all the coefficient matrices are 4×4 fuzzy matrices whose elements are in the form of TrFN given as follows:

$$\tilde{A} = \begin{bmatrix} 1 & (0.07, 0.15, 0.19, 0.27) & (-0.35, -0.30, -0.20, -0.15) & (0.50, 0.53, 0.56, 0.58) \\ (0.44, 0.46, 0.48, 0.50) & 1 & (0.65, 0.66, 0.68, 0.69) & (-0.34, -0.33, -0.31, -0.30) \\ (-0.15, -0.12, -0.10, -0.07) & (0.30, 0.33, 0.37, 0.40) & 1 & (-0.78, -0.75, -0.73, -0.70) \\ (0.50, 0.53, 0.57, 0.60) & (0.40, 0.41, 0.45, 0.46) & (0.30, 0.35, 0.37, 0.42) & 1 \end{bmatrix},$$

$$\tilde{B} = \begin{bmatrix} (0.20, 0.21, 0.23, 0.24) & (0.01, 0.015, 0.025, 0.03) & (0.10, 0.11, 0.13, 0.14) & (0.13, 0.135, 0.145, 0.15) \\ (0.01, 0.015, 0.025, 0.03) & (0.12, 0.13, 0.15, 0.16) & (0.03, 0.035, 0.045, 0.05) & (-0.08, -0.07, -0.05, -0.04) \\ (0.10, 0.11, 0.13, 0.14) & (0.03, 0.035, 0.045, 0.05) & (0.24, 0.027, 0.29, 0.32) & (0.04, 0.06, 0.10, 0.12) \\ (0.13, 0.135, 0.145, 0.15) & (-0.08, -0.07, -0.05, -0.04) & (0.04, 0.07, 0.09, 0.12) & (0.22, 0.24, 0.28, 0.30) \end{bmatrix},$$

$$\tilde{C} = \begin{bmatrix} -3.0475 & -2.1879 & -1.9449 & -2.8242 \\ -2.6500 & -2.4724 & -2.3515 & -2.1053 \\ -0.7456 & -0.6423 & -1.3117 & -0.1852 \\ -4.0500 & -3.0631 & -2.8121 & -3.7794 \end{bmatrix}.$$

7.9 REFERENCES

[1] Apel, T., Mehrmann, V., and Watkins, D., 2002. Structured eigenvalue methods for the computation of corner singularities in 3D anisotropic elastic structures. *Computer Methods in Applied Mechanics and Engineering*, 191(39–40):4459–4473. DOI: 10.1016/s0045-7825(02)00390-0. 126

[2] Bai, Z. and Su, Y., 2005. SOAR: A second-order Arnoldi method for the solution of the quadratic eigenvalue problem. *SIAM Journal on Matrix Analysis and Applications*, 26(3):640–659. DOI: 10.1137/s0895479803438523.

[3] Bender, C. M., Komijani, J., and Wang, Q. H., 2019. Nonlinear eigenvalue problems for generalized Painlevé equations. *Journal of Physics A: Mathematical and Theoretical*. DOI: 10.1088/1751-8121/ab2bcc.

[4] Chakraverty, S. and Mahato, N. R., 2018. Nonlinear interval eigenvalue problems for damped spring-mass system. *Engineering Computations*, 35(6):2272–2286. DOI: 10.1108/ec-04-2017-0128.

[5] Fazeli, S. A. and Rabiei, F., 2016. Solving nonlinear eigenvalue problems using an improved Newton method. *International Journal of Advanced Computer Science and Applications*, 7(9):438–441. DOI: 10.14569/ijacsa.2016.070959. 130

[6] Gao, W., Yang, C., and Meza, J. C., 2009. Solving a class of nonlinear eigenvalue problems by Newton's method (No. LBNL-2187E). Lawrence Berkeley National Lab. (LBNL), Berkeley, CA.

[7] Jeswal, S. K. and Chakraverty, S., 2019. Neural network approach for solving nonlinear eigenvalue problems of structural dynamics. *Neural Computing and Applications*, pages 1–9. DOI: 10.1007/s00521-019-04600-3.

[8] Kressner, D., 2009. A block Newton method for nonlinear eigenvalue problems. *Numerische Mathematik*, 114(2):355–372. DOI: 10.1007/s00211-009-0259-x.

[9] Kurseeva, V. Y., Tikhov, S. V., and Valovik, D. V., 2019. Nonlinear multiparameter eigenvalue problems: Linearised and nonlinearised solutions. *Journal of Differential Equations*, 267(4):2357–2384. DOI: 10.1016/j.jde.2019.03.014.

[10] Lawrence, P. W., Van Barel, M., and Van Dooren, P., 2016. Backward error analysis of polynomial eigenvalue problems solved by linearization. *SIAM Journal on Matrix Analysis and Applications*, 37(1):123–144. DOI: 10.1137/140979034.

[11] Leng, H., 2014. Real eigenvalue bounds of standard and generalized real interval eigenvalue problems. *Applied Mathematics and Computation*, 232:164–171. DOI: 10.1016/j.amc.2014.01.070.

[12] Mahato, N. R. and Chakraverty, S., 2016a. Filtering algorithm for real eigenvalue bounds of interval and fuzzy generalized eigenvalue problems. *ASCE-ASME Journal of Risk and Uncertainty in Engineering Systems, Part B: Mechanical Engineering*, 2(4):044502. DOI: 10.1115/1.4032958.

[13] Mahato, N. R. and Chakraverty, S., 2016b. Filtering algorithm for eigenvalue bounds of fuzzy symmetric matrices. *Engineering Computations*, 33(3):855–875. DOI: 10.1108/ec-12-2014-0255.

[14] Mehrmann, V. and Watkins, D., 2002. Polynomial eigenvalue problems with Hamiltonian structure. *Electronic Transactions on Numerical Analysis*, 13:106–118. 126, 127

[15] Mehrmann, V. and Watkins, D., 2002. Polynomial eigenvalue problems with Hamiltonian structure. *Electronic Transactions on Numerical Analysis*, 13:106–118.

[16] Rout, S. and Chakraverty, S., 2019. Solving fully fuzzy nonlinear eigenvalue problems of damped spring-mass structural systems using novel fuzzy-affine approach. *Computer Modeling in Engineering and Sciences*, 121(3):947–980. DOI: 10.32604/cmes.2019.08036.

[17] Rout, S. and Chakraverty, S., 2020. Affine approach to solve nonlinear eigenvalue problems of structures with uncertain parameters. In *Recent Trends in Wave Mechanics and Vibrations*. pages 407–425, Springer, Singapore. DOI: 10.1007/978-981-15-0287-3_29.

[18] Solovév, S. I. and Solovév, P. S., 2018. Finite element approximation of the minimal eigenvalue of a nonlinear eigenvalue problem. *Lobachevskii Journal of Mathematics*, 39(7):949–956. DOI: 10.1134/s199508021807020x.

[19] Su, Y. and Bai, Z., 2011. Solving rational eigenvalue problems via linearization. *SIAM Journal on Matrix Analysis and Applications*, 32(1):201–216. DOI: 10.1137/090777542.

[20] Tisseur, F., 2000. Backward error and condition of polynomial eigenvalue problems. *Linear Algebra and its Applications*, 309(1–3):339–361. DOI: 10.1016/s0024-3795(99)00063-4.

[21] Tisseur, F. and Meerbergen, K., 2001. The quadratic eigenvalue problem. *SIAM Review*, 43(2):235–286. DOI: 10.1137/s0036144500381988. 127

[22] Wetherhold, R. and Padliya, P. S., 2014. Design aspects of nonlinear vibration analysis of rectangular orthotropic membranes. *Journal of Vibration and Acoustics*, 136(3):034506. DOI: 10.1115/1.4027148.

Authors' Biographies

SNEHASHISH CHAKRAVERTY

Dr. Snehashish Chakraverty works in the Department of Mathematics (Applied Mathematics Group), National Institute of Technology Rourkela, Odisha, as a Senior (Higher Administrative Grade) Professor and is also the Dean of Student Welfare of the institute since November 2019. He received his Ph.D. from IIT Roorkee in 1992. Then he did post-doctoral research at ISVR, University of Southampton, U.K., and at Concordia University, Canada. He was a visiting professor at Concordia and McGill Universities, Canada, and University of Johannesburg, South Africa. Prof. Chakraverty has authored 17 books and published approximately 345 research papers in journals and conferences. He was the President of the Section of Mathematical Sciences (including Statistics) of Indian Science Congress (2015-2016) and was the Vice President—Orissa Mathematical Society (2011-2013). Prof. Chakraverty is a recipient of prestigious awards viz. INSA International Bilateral Exchange Program, Platinum Jubilee ISCA Lecture, CSIR Young Scientist, BOYSCAST, UCOST Young Scientist, Golden Jubilee CBRI Director's Award, Roorkee University gold Medals and more. He has undertaken 17 research projects as Principal Investigator funded by different agencies totaling about Rs.1.6 crores. Prof. Chakraverty is the Chief Editor of *International Journal of Fuzzy Computation and Modelling (IJFCM)*, Inderscience Publisher, Switzerland (`http://www.inderscience.com/ijfcm`), Associate Editor of *Computational Methods in Structural Engineering*, *Frontiers in Built Environment*, and an Editorial Board member of Springer Nature Applied Sciences, IGI Research Insights Books, Springer Book Series of Modeling and Optimization in Science and Technologies, Coupled Systems Mechanics (Techno Press), Curved and Layered Structures (De Gruyter), *Journal of Composites Science (MDPI)*, Engineering Research Express (IOP), *Applications and Applied Mathematics: An International Journal*, and Computational Engineering and Physical Modeling (Pouyan Press).

His present research area includes Differential Equations (Ordinary, Partial, and Fractional), Numerical Analysis and Computational Methods, Structural Dynamics (FGM, Nano), and Fluid Dynamics, Mathematical Modeling and Uncertainty Modeling, and Soft Computing and Machine Intelligence (Artificial Neural Network, Fuzzy, Interval, and Affine Computations).

SAUDAMINI ROUT

Saudamini Rout is currently pursuing her Ph.D. under the supervision of Prof. S. Chakraverty at the Department of Mathematics, National Institute of Technology Rourkela, Rourkela, Odisha, India. She completed her M.Phil. in Mathematics in 2016, M.Sc. in Mathematics in 2015, and B.Sc. in 2013 from Ravenshaw University, Cuttack, Odisha, India. She has been awarded a gold medal for securing first position in the university in M.Sc. She has been selected for the Biju Patnaik Research Fellowship, Department of Higher Education, Government of Odisha, India. She has participated in various conferences/workshops and published a few research articles and book chapters. She has also served as a reviewer for various international journals. Her current research interests include Fuzzy Set Theory, Interval Analysis, Affine Arithmetic, Uncertain Linear and/or Nonlinear Static and Dynamic Problems, and Numerical Analysis.

⊙编辑手记

A. Einstein 曾经说过:“所有科学理论应该尽量地简化,直到不能再简化为止”.这一点对数学理论来说尤其重要.

本书是一部试图简化版的模糊数学专著.

模糊数学(fuzzy mathematics)是研究和处理模糊性现象的数学理论和方法.1965年美国控制理论专家L.A.扎德发表了论文《模糊集合》(也有译成不分明集合,弗晰集合的),标志着这门新学科的诞生.

本书中文书名或可译为:《基于不确定静态和动态问题解的仿射算术》.本书的作者有两位:

一位是斯内哈希什·查克拉弗蒂(Snehashish Chakraverty),他是印度数学家,就职于奥里萨邦鲁尔克拉国立理工学院数学系(应用数学组),担任高级(高级行政级别)教授,自2019年11月起担任该学院学生福利部主任.他是许多著名奖项的获得者,比如INSA国际双边交流计划、白金禧年ISCA讲座、CSIR青年科学家、BOYSCAST、UCOST青年科学家、金禧CBRI主任奖、鲁尔基大学金奖等.他目前的研究领域包括微分方程、数值分析和计算方法、结构动力学(FGM、纳米)和流体动力学、数学建模和不确定性建模,以及软计算和机器智能(人工神经网络)等.

另一位绍达米尼·劳特(Saudamini Rout)也是印度数学家,她目前在印度奥里萨邦鲁尔克拉国立理工学院数学系的S.查克拉弗蒂教授的指导下攻读博士学位.她曾获得印度奥里萨邦政府高等教育部的Biju Patnaik研究奖学金.她参加过多种会议,并发表了一些研究文章和书籍章节.她还担任过各种国际期刊的审稿人.她目前的研究兴趣包括模糊集理论、区间分析、仿射算术、不确定线性和数值分析等.

模糊数学的产生是为了适应处理运用传统数学方法无法适当模拟的现象的迫切需要.因为这些现象或是很难给予确切定义,或是它们包含着实质上是模糊的相互关系,而具有不确定性的事物和现象不能用经典集合论来刻画.

正如作者在本书前言中所述:

静态问题和动态问题可以通过联立方程组、特征值问题、微分方程和积分方程组来建模.通常,这些问题的材料和几何特性(或涉及的参数)被认为是精确数值的形式.不确定性几乎与每一次测量都不可分割,在处理现实世界的问题时,不确定性的出现是必然的.因此,在实际实践中,由于维护、测量或实验引起的误差,材料特性可能是不确定的或模糊的.不确定模型有三种类型:概率或统计方法、区间分析、模糊集合论.在概率方法中,不确定参数被视为随机变量,而在区间分析和模糊计算中,参数分别被视为实数轴 **R** 上的闭区间和模糊数.从传统意义上来说,概率方法或统计方法被用来处理不确定性和模糊性.不幸的是,在没有足够的所涉参数的实验数据的情况下,概率方法可能不能以所需的精度提供可靠的结果.因此,区间分析和模糊集合论已成为近几十年来许多应用的有力工具.在模糊集合论和区间分析中,不确定参数通过模糊变量和区间变量来表达,比如模糊数和区间数,向量或模糊矩阵和区间矩阵.再进一步,模糊数可以被参数化和转化为区间的一个族.处理带区间分析的不确定性时的主要障碍是它的依赖性问题.区间计算假设计算的所有运算物体在执行任何区间操作时,在它们的范围内都是独立的.但是,当运算物体是部分独立时,区间计算可能导致区间解的宽度大于精确范围.在这方面,仿射算数(AA)可能是以不同方式处理不确定性的最新发展之一,这可能有助于克服依赖性问题并为解决方案计算更好的界线.进一步说,按照我们的目的,静态和动态问题分别转变成了线性/非线性方程组和特征值问题.同样地,仿射算数是基于解决以上问题和处理不确定性的方法而提出或发展出来的.

就这一点而言,本书将会由区间的基本前提和模糊算法开始,然后它将处理标准区间运算的依赖性问题和仿射算数的需要.穷举仿射算数运算和性质将会包括模糊一仿射算数.可以注意到带不确定性的静态问题导致了线性方程组的不确定性,而且不确定动态问题会转变为不确定性的线性和非线性特征值问题.

同样地,本书包含了处理之前描述的基于仿射和区间/模糊方法的问题的新开发的有效方法.我们研究了有关结构的静态和动态问题的各种说明性示例,以展示所有已开发方法的可靠性和有效性.作者假设读者具备微积分、实分析、线性代数、数值分析和微分方程的基本知识.

本书共包含7章,第1章包含了对书名所列问题的详细介绍和文献研究.第2章包括区间分析和模糊集合论的基本定义、术语和性质.第3章讨论了区间依赖性问题背后的原因和对仿射算数的详细的解释.为了有效地处理模糊数形式的带不确定性的现实生活中的问题,第4章提出了新的模糊一仿射算数.在第5章中,关于不确定静态问题的研究已经被合并了,其可能导致区间或模糊的线性方程组.带有不确定材料的动态分析和各种实际问题的几何性质可以通过区间或模糊特征值问题建模.这样一来,在第6章和第7章中分别展示了解决不确定线性和非线性动态问题的不同方法.在整本书中,不确定性以闭区间的形式呈现,而模糊性则以三角形和梯形模糊数的形式来考虑.

本书试图在不确定的环境中严谨地研究各种科学和工程领域的静态和动态问题.作者相信本书将为全球的本科生、研究生、研究人员、工业界、学院和其他人提供帮助.

模糊数学自1965年被创立以来,已经取得了巨大的发展,发表的相关论文的统计数字足以证明这点:

截止年限	论文及文献总数
1976	763
1980	约1 500
1982	2 410
1983	3 064
1984	约5 000

后面的年份就呈指数爆炸了!

本书的版权编辑李丹女士为了使读者能够更快了解本书,特翻译了本书目录如下:

2.4 区间集合运算

2.5 区间运算

2.6 区间代数性质

2.7 模糊集合

2.8 模糊集合的基本术语

2.9 模糊集合运算

2.10 模糊集合的算法

2.11 模糊数

2.12 模糊数的不同类型

2.13 练习

2.14 参考文献

3. 仿射算术

3.1 区间依赖性问题

3.2 定义与注释

3.3 仿射到区间的转换

3.4 区间到仿射的转换

3.5 仿射算术运算

3.6 仿射算术的作用

3.7 练习

3.8 参考文献

4. 模糊一仿射算术

4.1 模糊数的 a一剖分

4.2 模糊数的参数形式

4.3 参数模糊数的基本术语

4.4 参数模糊算术

4.5 模糊一仿射形式

4.6 模糊一仿射算术运算

4.7 模糊一仿射算术的作用

4.8 练习

4.9 参考文献

5. 不确定静态问题

5.1 克里斯普(Crisp)线性方程组(CSLE)

5.2 区间线性方程组(ISLE)

5.3 ISLE 的解

5.4 模糊线性方程组(FSLE)

5.5 FSLE 的解

5.6 练习

5.7 参考文献

6. 不确定线性动态问题

6.1 线性特征值问题(LEP)

6.2 克里斯普广义特征值问题(CGEP)

6.3 CGEP 的解

6.4 区间广义特征值问题(IGEP)

6.5 IGEP 的解

6.6 模糊广义特征值问题(FGEP)

6.7 FGEP 的解

6.8 练习

6.9 参考文献

7. 克里斯普非线性动态问题

7.1 克里斯普非线性特征值问题(CNEP)

7.2 CNEP 的线性化

7.3 CNEP 的解

7.4 区间非线性特征值问题(INEP)

7.5 INEP 的解

7.6 模糊非线性特征值问题(FNEP)

7.7 FNEP 的解

7.8 练习

7.9 参考文献

开尔文勋爵 1883 年曾说过:"在物理科学中,学习任何论题的关键的第一步是寻找它的数值计算原理和与之有关的一些性质的测量方法."模糊数学的开创者 L. A. 扎德在其《模糊集合、语言变量及模糊逻辑》一书的导言中写道:现代科学有一个基本信条.这个信条认为:对于一种现象,未经定理描述,就谈不上彻底理解.照此推论,构成科学知识的核心的大多数东西,可以看作一个概念和方法的贮水池,人们可以汲取这些概念和方法,为各种形式的系统建立数学模型,并通过这些数学模型获得关于系统特性的定量信息.

人们尊重精确、严格和定量的东西,看不起模糊、不严格和定性的东西,但有一个原理断言高精度与高复杂性是不兼容的,于是模糊数学开始登堂入市了!

今天是大年初一,已经是在疫情下度过的第三个春节了,昔日的流金岁月还会重现吗?William Somerset Maugham(1874—1965)曾说:

> 人们要为年轻时对未来的美好憧憬,付出饱尝幻灭之苦的惨痛代价.

刘培杰

2022 年 2 月 1 日

于哈工大

刘培杰数学工作室
已出版(即将出版)图书目录——原版影印

书　　名	出版时间	定　价	编号
数学物理大百科全书.第1卷(英文)	2016－01	418.00	508
数学物理大百科全书.第2卷(英文)	2016－01	408.00	509
数学物理大百科全书.第3卷(英文)	2016－01	396.00	510
数学物理大百科全书.第4卷(英文)	2016－01	408.00	511
数学物理大百科全书.第5卷(英文)	2016－01	368.00	512
zeta函数,q-zeta函数,相伴级数与积分(英文)	2015－08	88.00	513
微分形式:理论与练习(英文)	2015－08	58.00	514
离散与微分包含的逼近和优化(英文)	2015－08	58.00	515
艾伦·图灵:他的工作与影响(英文)	2016－01	98.00	560
测度理论概率导论,第2版(英文)	2016－01	88.00	561
带有潜在故障恢复系统的半马尔柯夫模型控制(英文)	2016－01	98.00	562
数学分析原理(英文)	2016－01	88.00	563
随机偏微分方程的有效动力学(英文)	2016－01	88.00	564
图的谱半径(英文)	2016－01	58.00	565
量子机器学习中数据挖掘的量子计算方法(英文)	2016－01	98.00	566
量子物理的非常规方法(英文)	2016－01	118.00	567
运输过程的统一非局部理论:广义波尔兹曼物理动力学,第2版(英文)	2016－01	198.00	568
量子力学与经典力学之间的联系在原子、分子及电动力学系统建模中的应用(英文)	2016－01	58.00	569
算术域(英文)	2018－01	158.00	821
高等数学竞赛:1962—1991年的米洛克斯·史怀哲竞赛(英文)	2018－01	128.00	822
用数学奥林匹克精神解决数论问题(英文)	2018－01	108.00	823
代数几何(德文)	2018－04	68.00	824
丢番图逼近论(英文)	2018－01	78.00	825
代数几何学基础教程(英文)	2018－01	98.00	826
解析数论入门课程(英文)	2018－01	78.00	827
数论中的丢番图问题(英文)	2018－01	78.00	829
数论(梦幻之旅):第五届中日数论研讨会演讲集(英文)	2018－01	68.00	830
数论新应用(英文)	2018－01	68.00	831
数论(英文)	2018－01	78.00	832

刘培杰数学工作室
已出版(即将出版)图书目录——原版影印

书　　名	出版时间	定　价	编号
湍流十讲(英文)	2018—04	108.00	886
无穷维李代数:第3版(英文)	2018—04	98.00	887
等值、不变量和对称性(英文)	2018—04	78.00	888
解析数论(英文)	2018—09	78.00	889
《数学原理》的演化:伯特兰·罗素撰写第二版时的手稿与笔记(英文)	2018—04	108.00	890
哈密尔顿数学论文集(第4卷):几何学、分析学、天文学、概率和有限差分等(英文)	2019—05	108.00	891
偏微分方程全局吸引子的特性(英文)	2018—09	108.00	979
整函数与下调和函数(英文)	2018—09	118.00	980
幂等分析(英文)	2018—09	118.00	981
李群,离散子群与不变量理论(英文)	2018—09	108.00	982
动力系统与统计力学(英文)	2018—09	118.00	983
表示论与动力系统(英文)	2018—09	118.00	984
分析学练习.第1部分(英文)	2021—01	88.00	1247
分析学练习.第2部分,非线性分析(英文)	2021—01	88.00	1248
初级统计学:循序渐进的方法:第10版(英文)	2019—05	68.00	1067
工程师与科学家微分方程用书:第4版(英文)	2019—07	58.00	1068
大学代数与三角学(英文)	2019—06	78.00	1069
培养数学能力的途径(英文)	2019—07	38.00	1070
工程师与科学家统计学:第4版(英文)	2019—06	58.00	1071
贸易与经济中的应用统计学:第6版(英文)	2019—06	58.00	1072
傅立叶级数和边值问题:第8版(英文)	2019—05	48.00	1073
通往天文学的途径:第5版(英文)	2019—05	58.00	1074
拉马努金笔记.第1卷(英文)	2019—06	165.00	1078
拉马努金笔记.第2卷(英文)	2019—06	165.00	1079
拉马努金笔记.第3卷(英文)	2019—06	165.00	1080
拉马努金笔记.第4卷(英文)	2019—06	165.00	1081
拉马努金笔记.第5卷(英文)	2019—06	165.00	1082
拉马努金遗失笔记.第1卷(英文)	2019—06	109.00	1083
拉马努金遗失笔记.第2卷(英文)	2019—06	109.00	1084
拉马努金遗失笔记.第3卷(英文)	2019—06	109.00	1085
拉马努金遗失笔记.第4卷(英文)	2019—06	109.00	1086
数论:1976年纽约洛克菲勒大学数论会议记录(英文)	2020—06	68.00	1145
数论:卡本代尔1979:1979年在南伊利诺伊卡本代尔大学举行的数论会议记录(英文)	2020—06	78.00	1146
数论:诺德韦克豪特1983:1983年在诺德韦克豪特举行的Journees Arithmetiques数论大会会议记录(英文)	2020—06	68.00	1147
数论:1985—1988年在纽约城市大学研究生院和大学中心举办的研讨会(英文)	2020—06	68.00	1148

刘培杰数学工作室
已出版(即将出版)图书目录——原版影印

书 名	出版时间	定 价	编号
数论:1987 年在乌尔姆举行的 Journees Arithmetiques 数论大会会议记录(英文)	2020—06	68.00	1149
数论:马德拉斯 1987:1987 年在马德拉斯安娜大学举行的国际拉马努金百年纪念大会会议记录(英文)	2020—06	68.00	1150
解析数论:1988 年在东京举行的日法研讨会会议记录(英文)	2020—06	68.00	1151
解析数论:2002 年在意大利切特拉罗举行的 C. I. M. E. 暑期班演讲集(英文)	2020—06	68.00	1152
量子世界中的蝴蝶:最迷人的量子分形故事(英文)	2020—06	118.00	1157
走进量子力学(英文)	2020—06	118.00	1158
计算物理学概论(英文)	2020—06	48.00	1159
物质,空间和时间的理论:量子理论(英文)	2020—10	48.00	1160
物质,空间和时间的理论:经典理论(英文)	2020—10	48.00	1161
量子场理论:解释世界的神秘背景(英文)	2020—07	38.00	1162
计算物理学概论(英文)	2020—06	48.00	1163
行星状星云(英文)	2020—10	38.00	1164
基本宇宙学:从亚里士多德的宇宙到大爆炸(英文)	2020—08	58.00	1165
数学磁流体力学(英文)	2020—07	58.00	1166
计算科学:第 1 卷,计算的科学(日文)	2020—07	88.00	1167
计算科学:第 2 卷,计算与宇宙(日文)	2020—07	88.00	1168
计算科学:第 3 卷,计算与物质(日文)	2020—07	88.00	1169
计算科学:第 4 卷,计算与生命(日文)	2020—07	88.00	1170
计算科学:第 5 卷,计算与地球环境(日文)	2020—07	88.00	1171
计算科学:第 6 卷,计算与社会(日文)	2020—07	88.00	1172
计算科学. 别卷,超级计算机(日文)	2020—07	88.00	1173
多复变函数论(日文)	2022—06	78.00	1518
复变函数入门(日文)	2022—06	78.00	1523
代数与数论:综合方法(英文)	2020—10	78.00	1185
复分析:现代函数理论第一课(英文)	2020—07	58.00	1186
斐波那契数列和卡特兰数:导论(英文)	2020—10	68.00	1187
组合推理:计数艺术介绍(英文)	2020—07	88.00	1188
二次互反律的傅里叶分析证明(英文)	2020—07	48.00	1189
旋瓦兹分布的希尔伯特变换与应用(英文)	2020—07	58.00	1190
泛函分析:巴拿赫空间理论入门(英文)	2020—07	48.00	1191
卡塔兰数入门(英文)	2019—05	68.00	1060
测度与积分(英文)	2019—04	68.00	1059
组合学手册. 第一卷(英文)	2020—06	128.00	1153
*一代数、局部紧群和巴拿赫*一代数丛的表示. 第一卷,群和代数的基本表示理论(英文)	2020—05	148.00	1154
电磁理论(英文)	2020—08	48.00	1193
连续介质力学中的非线性问题(英文)	2020—09	78.00	1195
多变量数学入门(英文)	2021—05	68.00	1317
偏微分方程入门(英文)	2021—05	88.00	1318
若尔当典范性:理论与实践(英文)	2021—07	68.00	1366
伽罗瓦理论. 第 4 版(英文)	2021—08	88.00	1408
R 统计学概论	2023—03	88.00	1614
基于不确定静态和动态问题解的仿射算术(英文)	2023—03	38.00	1618

刘培杰数学工作室
已出版(即将出版)图书目录——原版影印

书　　名	出版时间	定　价	编号
典型群,错排与素数(英文)	2020－11	58.00	1204
李代数的表示:通过 gln 进行介绍(英文)	2020－10	38.00	1205
实分析演讲集(英文)	2020－10	38.00	1206
现代分析及其应用的课程(英文)	2020－10	58.00	1207
运动中的抛射物数学(英文)	2020－10	38.00	1208
2－纽结与它们的群(英文)	2020－10	38.00	1209
概率,策略和选择:博弈与选举中的数学(英文)	2020－11	58.00	1210
分析学引论(英文)	2020－11	58.00	1211
量子群:通往流代数的路径(英文)	2020－11	38.00	1212
集合论入门(英文)	2020－10	48.00	1213
酉反射群(英文)	2020－11	58.00	1214
探索数学:吸引人的证明方式(英文)	2020－11	58.00	1215
微分拓扑短期课程(英文)	2020－10	48.00	1216
抽象凸分析(英文)	2020－11	68.00	1222
费马大定理笔记(英文)	2021－03	48.00	1223
高斯与雅可比和(英文)	2021－03	78.00	1224
π与算术几何平均:关于解析数论和计算复杂性的研究(英文)	2021－01	58.00	1225
复分析入门(英文)	2021－03	48.00	1226
爱德华・卢卡斯与素性测定(英文)	2021－03	78.00	1227
通往凸分析及其应用的简单路径(英文)	2021－01	68.00	1229
微分几何的各个方面.第一卷(英文)	2021－01	58.00	1230
微分几何的各个方面.第二卷(英文)	2020－12	58.00	1231
微分几何的各个方面.第三卷(英文)	2020－12	58.00	1232
沃克流形几何学(英文)	2020－11	58.00	1233
彷射和韦尔几何应用(英文)	2020－12	58.00	1234
双曲几何学的旋转向量空间方法(英文)	2021－02	58.00	1235
积分:分析学的关键(英文)	2020－12	48.00	1236
为有天分的新生准备的分析学基础教材(英文)	2020－11	48.00	1237
数学不等式.第一卷.对称多项式不等式(英文)	2021－03	108.00	1273
数学不等式.第二卷.对称有理不等式与对称无理不等式(英文)	2021－03	108.00	1274
数学不等式.第三卷.循环不等式与非循环不等式(英文)	2021－03	108.00	1275
数学不等式.第四卷.Jensen 不等式的扩展与加细(英文)	2021－03	108.00	1276
数学不等式.第五卷.创建不等式与解不等式的其他方法(英文)	2021－04	108.00	1277

刘培杰数学工作室
已出版(即将出版)图书目录——原版影印

书　　名	出版时间	定　价	编号
冯·诺依曼代数中的谱位移函数:半有限冯·诺依曼代数中的谱位移函数与谱流(英文)	2021—06	98.00	1308
链接结构:关于嵌入完全图的直线中链接单形的组合结构(英文)	2021—05	58.00	1309
代数几何方法.第1卷(英文)	2021—06	68.00	1310
代数几何方法.第2卷(英文)	2021—06	68.00	1311
代数几何方法.第3卷(英文)	2021—06	58.00	1312

书　　名	出版时间	定　价	编号
代数、生物信息和机器人技术的算法问题.第四卷,独立恒等式系统(俄文)	2020—08	118.00	1199
代数、生物信息和机器人技术的算法问题.第五卷,相对覆盖性和独立可拆分恒等式系统(俄文)	2020—08	118.00	1200
代数、生物信息和机器人技术的算法问题.第六卷,恒等式和准恒等式的相等 问题、可推导性和可实现性(俄文)	2020—08	128.00	1201
分数阶微积分的应用:非局部动态过程,分数阶导热系数(俄文)	2021—01	68.00	1241
泛函分析问题与练习:第2版(俄文)	2021—01	98.00	1242
集合论、数学逻辑和算法论问题:第5版(俄文)	2021—01	98.00	1243
微分几何和拓扑短期课程(俄文)	2021—01	98.00	1244
素数规律(俄文)	2021—01	88.00	1245
无穷边值问题解的递减:无界域中的拟线性椭圆和抛物方程(俄文)	2021—01	48.00	1246
微分几何讲义(俄文)	2020—12	98.00	1253
二次型和矩阵(俄文)	2021—01	98.00	1255
积分和级数.第2卷,特殊函数(俄文)	2021—01	168.00	1258
积分和级数.第3卷,特殊函数补充:第2版(俄文)	2021—01	178.00	1264
几何图上的微分方程(俄文)	2021—01	138.00	1259
数论教程:第2版(俄文)	2021—01	98.00	1260
非阿基米德分析及其应用(俄文)	2021—03	98.00	1261
古典群和量子群的压缩(俄文)	2021—03	98.00	1263
数学分析习题集.第3卷,多元函数:第3版(俄文)	2021—03	98.00	1266
数学习题:乌拉尔国立大学数学力学系大学生奥林匹克(俄文)	2021—03	98.00	1267
柯西定理和微分方程的特解(俄文)	2021—03	98.00	1268
组合极值问题及其应用:第3版(俄文)	2021—03	98.00	1269
数学词典(俄文)	2021—01	98.00	1271
确定性混沌分析模型(俄文)	2021—06	168.00	1307
精选初等数学习题和定理.立体几何.第3版(俄文)	2021—03	68.00	1316
微分几何习题:第3版(俄文)	2021—05	98.00	1336
精选初等数学习题和定理.平面几何.第4版(俄文)	2021—05	68.00	1335
曲面理论在欧氏空间 En 中的直接表示(俄文)	2022—01	68.00	1444
维纳一霍普夫离散算子和托普利兹算子:某些可数赋范空间中的诺特性和可逆性(俄文)	2022—03	108.00	1496
Maple 中的数论:数论中的计算机计算(俄文)	2022—03	88.00	1497
贝尔曼和克努特问题及其概括:加法运算的复杂性(俄文)	2022—03	138.00	1498

刘培杰数学工作室
已出版(即将出版)图书目录——原版影印

书　名	出版时间	定　价	编号
复分析:共形映射(俄文)	2022－07	48.00	1542
微积分代数样条和多项式及其在数值方法中的应用(俄文)	2022－08	128.00	1543
蒙特卡罗方法中的随机过程和场模型:算法和应用(俄文)	2022－08	88.00	1544
线性椭圆型方程组:论二阶椭圆型方程的迪利克雷问题(俄文)	2022－08	98.00	1561
动态系统解的增长特性:估值、稳定性、应用(俄文)	2022－08	118.00	1565
群的自由积分解:建立和应用(俄文)	2022－08	78.00	1570
混合方程和偏差自变数方程问题:解的存在和唯一性(俄文)	2023－01	78.00	1582
拟度量空间分析:存在和逼近定理(俄文)	2023－01	108.00	1583
二维和三维流形上函数的拓扑性质:函数的拓扑分类(俄文)	2023－03	68.00	1584
齐次马尔科夫过程建模的矩阵方法:此类方法能够用于不同目上的的复杂系统研究、设计和完善(俄文)	2023－03	68.00	1594
周期函数的近似方法和特性:特殊课程(俄文)	2023－04	158.00	1622
扩散方程解的矩函数:变分法(俄文)	2023－03	58.00	1623

书　名	出版时间	定　价	编号
狭义相对论与广义相对论:时空与引力导论(英文)	2021－07	88.00	1319
束流物理学和粒子加速器的实践介绍:第2版(英文)	2021－07	88.00	1320
凝聚态物理中的拓扑和微分几何简介(英文)	2021－05	88.00	1321
混沌映射:动力学、分形学和快速涨落(英文)	2021－05	128.00	1322
广义相对论:黑洞、引力波和宇宙学介绍(英文)	2021－06	68.00	1323
现代分析电磁均质化(英文)	2021－06	68.00	1324
为科学家提供的基本流体动力学(英文)	2021－06	88.00	1325
视觉天文学:理解夜空的指南(英文)	2021－06	68.00	1326
物理学中的计算方法(英文)	2021－06	68.00	1327
单星的结构与演化:导论(英文)	2021－06	108.00	1328
超越居里:1903年至1963年物理界四位女性及其著名发现(英文)	2021－06	68.00	1329
范德瓦尔斯流体热力学的进展(英文)	2021－06	68.00	1330
先进的托卡马克稳定性理论(英文)	2021－06	88.00	1331
经典场论导论:基本相互作用的过程(英文)	2021－07	88.00	1332
光致电离量子动力学方法原理(英文)	2021－07	108.00	1333
经典域论和应力:能量张量(英文)	2021－05	88.00	1334
非线性太赫兹光谱的概念与应用(英文)	2021－06	68.00	1337
电磁学中的无穷空间并矢格林函数(英文)	2021－06	88.00	1338
物理科学基础数学.第1卷,齐次边值问题、傅里叶方法和特殊函数(英文)	2021－07	108.00	1339
离散量子力学(英文)	2021－07	68.00	1340
核磁共振的物理学和数学(英文)	2021－07	108.00	1341
分子水平的静电学(英文)	2021－08	68.00	1342
非线性波:理论、计算机模拟、实验(英文)	2021－06	108.00	1343
石墨烯光学:经典问题的电解解决方案(英文)	2021－06	68.00	1344
超材料多元宇宙(英文)	2021－07	68.00	1345
银河系外的天体物理学(英文)	2021－07	68.00	1346
原子物理学(英文)	2021－07	68.00	1347
将光打结:将拓扑学应用于光学(英文)	2021－07	68.00	1348
电磁学:问题与解法(英文)	2021－07	88.00	1364
海浪的原理:介绍量子力学的技巧与应用(英文)	2021－07	108.00	1365
多孔介质中的流体:输运与相变(英文)	2021－07	68.00	1372
洛伦兹群的物理学(英文)	2021－08	68.00	1373
物理导论的数学方法和解决方法手册(英文)	2021－08	68.00	1374

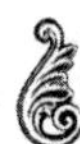

刘培杰数学工作室

已出版(即将出版)图书目录——原版影印

书　　名	出版时间	定　价	编号
非线性波数学物理学入门(英文)	2021—08	88.00	1376
波:基本原理和动力学(英文)	2021—07	68.00	1377
光电子量子计量学.第1卷,基础(英文)	2021—07	88.00	1383
光电子量子计量学.第2卷,应用与进展(英文)	2021—07	68.00	1384
复杂流的格子玻尔兹曼建模的工程应用(英文)	2021—08	68.00	1393
电偶极矩挑战(英文)	2021—08	108.00	1394
电动力学:问题与解法(英文)	2021—09	68.00	1395
自由电子激光的经典理论(英文)	2021—08	68.00	1397
曼哈顿计划——核武器物理学简介(英文)	2021—09	68.00	1401
粒子物理学(英文)	2021—09	68.00	1402
引力场中的量子信息(英文)	2021—09	128.00	1403
器件物理学的基本经典力学(英文)	2021—09	68.00	1404
等离子体物理及其空间应用导论.第1卷,基本原理和初步过程(英文)	2021—09	68.00	1405
磁约束聚变等离子体物理:理想 MHD 理论(英文)	2023—03	68.00	1613
相对论量子场论.第1卷,典范形式体系(英文)	2023—03	38.00	1615
涌现的物理学(英文)	2023—05	58.00	1619
量子化旋涡:一本拓扑激发手册(英文)	2023—04	68.00	1620
非线性动力学:实践的介绍性调查(英文)	2023—05	68.00	1621
拓扑与超弦理论焦点问题(英文)	2021—07	58.00	1349
应用数学:理论、方法与实践(英文)	2021—07	78.00	1350
非线性特征值问题:牛顿型方法与非线性瑞利函数(英文)	2021—07	58.00	1351
广义膨胀和齐性:利用齐性构造齐次系统的李雅普诺夫函数和控制律(英文)	2021—06	48.00	1352
解析数论焦点问题(英文)	2021—07	58.00	1353
随机微分方程:动态系统方法(英文)	2021—07	58.00	1354
经典力学与微分几何(英文)	2021—07	58.00	1355
负定相交形式流形上的瞬子模空间几何(英文)	2021—07	68.00	1356
广义卡塔兰轨道分析:广义卡塔兰轨道计算数字的方法(英文)	2021—07	48.00	1367
洛伦兹方法的变分:二维与三维洛伦兹方法(英文)	2021—08	38.00	1378
几何、分析和数论精编(英文)	2021—08	68.00	1380
从一个新角度看数论:通过遗传方法引入现实的概念(英文)	2021—07	58.00	1387
动力系统:短期课程(英文)	2021—08	68.00	1382
几何路径:理论与实践(英文)	2021—08	48.00	1385
论天体力学中某些问题的不可积性(英文)	2021—07	88.00	1396
广义斐波那契数列及其性质(英文)	2021—08	38.00	1386
对称函数和麦克唐纳多项式:余代数结构与 Kawanaka 恒等式(英文)	2021—09	38.00	1400
杰弗里·英格拉姆·泰勒科学论文集:第1卷.固体力学(英文)	2021—05	78.00	1360
杰弗里·英格拉姆·泰勒科学论文集:第2卷.气象学、海洋学和湍流(英文)	2021—05	68.00	1361
杰弗里·英格拉姆·泰勒科学论文集:第3卷.空气动力学以及落弹数和爆炸的力学(英文)	2021—05	68.00	1362
杰弗里·英格拉姆·泰勒科学论文集:第4卷.有关流体力学(英文)	2021—05	58.00	1363

刘培杰数学工作室

已出版(即将出版)图书目录——原版影印

书　　名	出版时间	定　价	编号
非局域泛函演化方程:积分与分数阶(英文)	2021—08	48.00	1390
理论工作者的高等微分几何:纤维丛、射流流形和拉格朗日理论(英文)	2021—08	68.00	1391
半线性退化椭圆微分方程:局部定理与整体定理(英文)	2021—07	48.00	1392
非交换几何、规范理论和重整化:一般简介与非交换量子场论的重整化(英文)	2021—09	78.00	1406
数论论文集:拉普拉斯变换和带有数论系数的幂级数(俄文)	2021—09	48.00	1407
挠理论专题:相对极大值,单射与扩充模(英文)	2021—09	88.00	1410
强正则图与欧几里得若尔当代数:非通常关系中的启示(英文)	2021—10	48.00	1411
拉格朗日几何和哈密顿几何:力学的应用(英文)	2021—10	48.00	1412
时滞微分方程与差分方程的振动理论:二阶与三阶(英文)	2021—10	98.00	1417
卷积结构与几何函数理论:用以研究特定几何函数理论方向的分数阶微积分算子与卷积结构(英文)	2021—10	48.00	1418
经典数学物理的历史发展(英文)	2021—10	78.00	1419
扩展线性丢番图问题(英文)	2021—10	38.00	1420
一类混沌动力系统的分歧分析与控制:分歧分析与控制(英文)	2021—11	38.00	1421
伽利略空间和伪伽利略空间中一些特殊曲线的几何性质(英文)	2022—01	68.00	1422
一阶偏微分方程:哈密尔顿—雅可比理论(英文)	2021—11	48.00	1424
各向异性黎曼多面体的反问题:分段光滑的各向异性黎曼多面体反边界谱问题:唯一性(英文)	2021—11	38.00	1425
项目反应理论手册.第一卷,模型(英文)	2021—11	138.00	1431
项目反应理论手册.第二卷,统计工具(英文)	2021—11	118.00	1432
项目反应理论手册.第三卷,应用(英文)	2021—11	138.00	1433
二次无理数:经典数论入门(英文)	2022—05	138.00	1434
数,形与对称性:数论,几何和群论导论(英文)	2022—05	128.00	1435
有限域手册(英文)	2021—11	178.00	1436
计算数论(英文)	2021—11	148.00	1437
拟群与其表示简介(英文)	2021—11	88.00	1438
数论与密码学导论:第二版(英文)	2022—01	148.00	1423

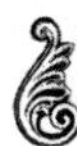 刘培杰数学工作室 已出版(即将出版)图书目录——原版影印 

书名	出版时间	定价	编号
几何分析中的柯西变换与黎兹变换:解析调和容量和李普希兹调和容量、变化和振荡以及一致可求长性(英文)	2021—12	38.00	1465
近似不动点定理及其应用(英文)	2022—05	28.00	1466
局部域的相关内容解析:对局部域的扩展及其伽罗瓦群的研究(英文)	2022—01	38.00	1467
反问题的二进制恢复方法(英文)	2022—03	28.00	1468
对几何函数中某些类的各个方面的研究:复变量理论(英文)	2022—01	38.00	1469
覆盖、对应和非交换几何(英文)	2022—01	28.00	1470
最优控制理论中的随机线性调节器问题:随机最优线性调节器问题(英文)	2022—01	38.00	1473
正交分解法:涡流流体动力学应用的正交分解法(英文)	2022—01	38.00	1475
芬斯勒几何的某些问题(英文)	2022—03	38.00	1476
受限三体问题(英文)	2022—05	38.00	1477
利用马利亚万微积分进行 Greeks 的计算:连续过程、跳跃过程中的马利亚万微积分和金融领域中的 Greeks(英文)	2022—05	48.00	1478
经典分析和泛函分析的应用:分析学的应用(英文)	2022—03	38.00	1479
特殊芬斯勒空间的探究(英文)	2022—03	48.00	1480
某些图形的施泰纳距离的细谷多项式:细谷多项式与图的维纳指数(英文)	2022—05	38.00	1481
图论问题的遗传算法:在新鲜与模糊的环境中(英文)	2022—05	48.00	1482
多项式映射的渐近簇(英文)	2022—05	38.00	1483
一维系统中的混沌:符号动力学,映射序列,一致收敛和沙可夫斯基定理(英文)	2022—05	38.00	1509
多维边界层流动与传热分析:粘性流体流动的数学建模与分析(英文)	2022—05	38.00	1510
演绎理论物理学的原理:一种基于量子力学波函数的逐次置信估计的一般理论的提议(英文)	2022—05	38.00	1511
R^2 和 R^3 中的仿射弹性曲线:概念和方法(英文)	2022—08	38.00	1512
算术数列中除数函数的分布:基本内容、调查、方法、第二矩、新结果(英文)	2022—05	28.00	1513
抛物型狄拉克算子和薛定谔方程:不定常薛定谔方程的抛物型狄拉克算子及其应用(英文)	2022—07	28.00	1514
黎曼-希尔伯特问题与量子场论:可积重正化、戴森-施温格方程(英文)	2022—08	38.00	1515
代数结构和几何结构的形变理论(英文)	2022—08	48.00	1516
概率结构和模糊结构上的不动点:概率结构和直觉模糊度量空间的不动点定理(英文)	2022—08	38.00	1517

刘培杰数学工作室
已出版(即将出版)图书目录——原版影印

书　　名	出版时间	定　价	编号
反若尔当对:简单反若尔当对的自同构(英文)	2022—07	28.00	1533
对某些黎曼—芬斯勒空间变换的研究:芬斯勒几何中的某些变换(英文)	2022—07	38.00	1534
内诣零流形映射的尼尔森数的阿诺索夫关系(英文)	2023—01	38.00	1535
与广义积分变换有关的分数次演算:对分数次演算的研究(英文)	2023—01	48.00	1536
强子的芬斯勒几何和吕拉几何(宇宙学方面):强子结构的芬斯勒几何和吕拉几何(拓扑缺陷)(英文)	2022—08	38.00	1537
一种基于混沌的非线性最优化问题:作业调度问题(英文)	2023—03	38.00	1538
广义概率论发展前景:关于趣味数学与置信函数实际应用的一些原创观点(英文)	2023—03	48.00	1539
纽结与物理学:第二版(英文)	2022—09	118.00	1547
正交多项式和q—级数的前沿(英文)	2022—09	98.00	1548
算子理论问题集(英文)	2022—09	108.00	1549
抽象代数:群、环与域的应用导论:第二版(英文)	2023—01	98.00	1550
菲尔兹奖得主演讲集:第三版(英文)	2023—01	138.00	1551
多元实函数教程(英文)	2022—09	118.00	1552
球面空间形式群的几何学:第二版(英文)	2022—09	98.00	1566
对称群的表示论(英文)	2023—01	98.00	1585
纽结理论:第二版(英文)	2023—01	88.00	1586
拟群理论的基础与应用(英文)	2023—01	88.00	1587
组合学:第二版(英文)	2023—01	98.00	1588
加性组合学:研究问题手册(英文)	2023—01	68.00	1589
扭曲、平铺与镶嵌:几何折纸中的数学方法(英文)	2023—01	98.00	1590
离散与计算几何手册:第三版(英文)	2023—01	248.00	1591
离散与组合数学手册:第二版(英文)	2023—01	248.00	1592
分析学教程.第1卷,一元实变量函数的微积分分析学介绍(英文)	2023—01	118.00	1595
分析学教程.第2卷,多元函数的微分和积分,向量微积分(英文)	2023—01	118.00	1596
分析学教程.第3卷,测度与积分理论,复变量的复值函数(英文)	2023—01	118.00	1597
分析学教程.第4卷,傅里叶分析,常微分方程,变分法(英文)	2023—01	118.00	1598

联系地址:哈尔滨市南岗区复华四道街10号　**哈尔滨工业大学出版社刘培杰数学工作室**

网　　址:http://lpj.hit.edu.cn/

邮　　编:150006

联系电话:0451—86281378　　13904613167

E-mail:lpj1378@163.com